Collins *gem*

Space Exploration

David Hawksett
Consultant: Dr Duncan Copp

Collins

An Imprint of HarperCollinsPublishers

ISBN-10: 0-00-710142-2
ISBN-13: 978-0-00-710142-9

ISBN-10: 0-06-081874-3 (in the United States)
ISBN-13: 978-0-06-081874-6

FIRST U.S. EDITION.

Printed and bound in Italy by Amadeus
10 09 08 07 06 05
9 8 7 6 5 4 3 2 1

Contents

Introduction

SPACE HAS ALWAYS played a part in our lives, from our
earliest ancestors who looked up at objects in the sky with
wonder, to the present generation. To our ancestors, the Sun,
the Moon and the stars in the night sky were things to be
worshipped, and even feared. The story of man's discovery
of space is one of the great epics in human history.

The rate of scientific discovery increased dramatically in the
twentieth century. We discovered other galaxies, how stars are
born and die, and the awesome power of black holes, which
suck up passing stars like cosmic vacuum cleaners. We are even
close to answering one of the most fundamental questions:
where did the Universe come from and what is its fate?

EXPLORING SPACE

WITH TELESCOPES and spacecraft, our exploration of the
heavens has made our world seem smaller, and the Universe
much larger. The planets are now more than mere points of
light in the sky – they are real places, visited by space probes.
Humans have left the Earth and entered space, and 12 men
have walked on the Moon.

Today, a multitude of satellites orbits the Earth, monitoring
our weather and environment, as well as providing global
communication. In the last few years, astronomers have
discovered dozens of planets around other stars. When they
study these new worlds in detail, perhaps we shall discover
whether or not we are alone in the Universe.

How To Use This Book

*C*OLLINS GEM *Space Exploration* will tell you all you ever wanted to know about space and space exploration. Divided into 10 sections, it will introduce you to the origins and contents of the Universe, before looking at how man's fascination with the night sky led to the era of space exploration.

Part One takes a look at what space is and the many explanations that have been given to try and explain its beginnings, including the Big Bang theory. Part Two is concerned with how we on Earth view what is out there in space, starting with the first telescopes through to the most modern techniques such as astrophotography. Part Three peers deep into space and explains how people have tried to map and measure the heavens, from primitive star maps drawn on cave walls to complex computer-generated star charts, and looks at whether or not the current expansion of the Universe will continue forever.

Part Four gives an overview of how space affects life on Earth: gravity, eclipses and metorites all play important roles, as do the presence of satellites, and technology first developed for use on space missions is now an integral part of life on our planet. Part Five looks at what is actually out there: the planets, stars and galaxies that we believe make up our Universe.

Part Six follows the journeys of the rockets, satellites and probes that have been sent into space to discover more of its mysteries. Parts Seven and Eight give an overview of life in space: the first human in space, the missions to the Moon, and how astronauts and cosmonauts survive while they are up there,

A

B

C

E

D

either in space stations or spacecraft. Part Nine is concerned
with the future of space exploration – where will we go next and
how long can we continue to send people on these missions?

Part Ten is a Compendium of fascinating facts: how to start
stargazing, planetary data, information on spacecraft missions,
further reading and useful websites, and a helpful glossary
and index.

A The page number appears in a colour-coded box,
indicating which section you are looking at.

B The aspect of space exploration you are looking at is
indicated at the head of the appropriate page.

C The text covers all aspects of space and space exploration.

D Biographical boxes on some spreads give details
of people who have shaped and influenced the history
of space exploration.

E Photographs and illustrations work with the text to
give an all-round view of space and space exploration.

Timeline

2136 BC Chinese record an eclipse of the Sun

350 BC Aristotle proposes the Earth is round

134 BC Hipparchus invents 'magnitude scale' for comparing star brightnesses

1054 Chinese observe supernova in the constellation of Taurus

1543 Nicholas Copernicus publishes his theory of the Sun-centred Solar System

1609 Galileo Galilei constructs his first astronomical refracting telescope and starts observing the skies

1610 Johannes Kepler argues that the Universe is not infinite; Galileo Galilei observes Jupiter's moons Io, Europa, Ganymede and Callisto, and first sees Saturn's rings without realising their nature

1613 Galileo Galilei discovers the Sun rotates by observing sunspots

1655 Giovanni Cassini discovers Jupiter's Great Red Spot

1656 Christiaan Huygens explains Saturn's rings and discovers the moon Titan

1668 Isaac Newton builds first reflecting telescope

1675 Ole Römer uses Jupiter's moons to calculate the speed of light

1705 Edmund Halley successfully predicts the return of a comet in 1758

The original telescope through which William Herschel discovered Uranus in 1781.

1781 William Herschel discovers Uranus

1784	John Mitchell proposes astronomical bodies from which light cannot escape – black holes
1796	Pierre Laplace proposes formation of the Solar System from a cloud of gas and dust
1801	Giuseppe Piazzi discovers the first asteroid, Ceres
1845	Third Earl of Rosse discovers the spiral nature of galaxies
1846	Neptune discovered after independent predictions by John Couch-Adams and Urbain Leverrier; William Lassell discovers Triton
1864	William Huggins discovers the Orion Nebula to be a gas cloud in space
1866	Giovanni Schiaparelli proposes meteor showers come from comet's tails
1910	Ejnar Hertzsprung and Henry Norris Russell find relationship between the colour and luminosity of the stars
1929	Edwin Hubble proves the expansion of the Universe, using the redshifts of galaxies; George Gamow discovers hydrogen fusion to be the energy supply for stars
1930	Clyde Tombaugh discovers Pluto
1932	Karl Jansky observes natural radio waves coming from the galactic centre
1942	Astronomers realise the Crab Nebula is debris from the 1054 supernova
1949	Herbert Friedman discovers X-rays from the Sun
1957	Jodrell Bank radio telescope completed near Manchester; Sputnik 1 becomes first artificial satellite
1961	Yuri Gagarin becomes first human in space
1964	Arno Penzias and Robert Wilson discover the cosmic microwave background radiation
1965	Penzias and Wilson publish details of their discovery;

Mariner 4 reaches Mars; Alexei Leonov performs
first spacewalk

1967   Jocelyn Bell and Antony Hewish discover the first pulsar
– created by the 1054 supernova

1969   Apollo 11 lands on the Moon and the first humans
walk on the lunar surface

1970   Venera 7 touches down on Venus

1971   Salyut 1 becomes first space station

1974   Mariner 10 becomes the only spacecraft to visit Mercury

1976   Vikings 1 and 2 touch down safely on Mars

1977   James Elliot discovers rings of Uranus; Voyagers 1 and 2
launched to the outer Solar System

1979   Voyager discovers active volcanoes on Jupiter's moon, Io

Buzz Aldrin was the second man on the Moon in July 1969.

An artist's impression of how the International Space Station will look like when it is completed.

1986 Launch of Mir space station; Giotto probe encounters Halley's Comet

1987 Supernova observed in the Large Magellanic Cloud

1990 Hubble Space Telescope launched; Magellan spacecraft begins four-year orbital mapping of Venus

1992 COBE satellite finds seeds of galaxies in the cosmic microwave background radiation

1995 Galileo spacecraft begins orbital survey of Jupiter and its moons

1997 Pathfinder releases first rover on to the Martian surface

2000 First astronauts live on board the International Space Station (ISS)

2004 Cassini probe begins exploration of Saturn system

2005 MERs (Mars Exploration Rovers) complete one Earth year on Mars

UNDERSTANDING SPACE

Introduction

ONE OF MAN'S basic characteristics is the urge to explore his world, including the wondrous events he witnesses in the sky. The quest to understand why the Sun rises and sets, why the Moon changes from night to night, and what the stars are, has been constant since prehistory.

Astronomy was of great significance to ancient cultures. The three most important centres of ancient astronomy were China, Egypt and Chaldea in the plains of Mesopotamia. Common to all these cultures was the belief that the heavenly bodies could help them predict the future.

ANCIENT OBSERVATIONS

MANY OF THE observations made by the ancient cultures were incredibly accurate. The Egyptians built their pyramids aligned to within a few tenths of a degree of the four points of the compass, by observing the stars. The Chinese recorded important astronomical events such as eclipses and supernovae. An observation of the first five planets to be discovered is thought to have been as early as 2449 BC.

However, sophisticated though these observations were, the ancient civilisations still believed the Earth was flat and that the hollow dome of the sky above was held in place by pillars.

We now know, of course, that the Earth is round, and with telescopes and space probes we have been able to see other planets close up and observe distant galaxies. Defining space itself is harder: the more scientists discover, the more questions there are to be asked. Is space infinite? If not, where does it start and end? One simple way of describing it is the near vacuum beyond our atmosphere in which all the objects in the Universe, including Earth, exist. But we still have a long way to go until we fully understand what 'space' is.

The Sun has been a source of fascination for many millennia.

Early View of Space

OUR PERCEPTION of the Universe began to change in the sixth century BC, with the advent of Greek astronomy. Instead of just observing the Universe, the Greeks attempted to explain rationally what they were seeing, rather than rely purely on superstition and religion, and astronomy as a true science was born.

Ptolemy

Ptolemy (c. AD 120–180) lived and worked in Alexandria, Egypt. He is best remembered for writing *Almagest*, in which he attempted to explain the motions of the Sun, Moon and planets against the backdrop of the stars. His theory, in which the Earth was the centre of the Universe, dominated astronomy until the sixteenth century.

THE GREEKS

IN THE fourth century BC, Plato (427–347 BC) described his theory of a finite enclosed Universe, floating in an infinite space. The Earth was supreme at the centre and the Sun, Moon, stars and planets revolved around it.

The philosopher Aristotle (384–322 BC) was the first to show that the Earth must be spherical, by observing the curved shadow of the Earth on the Moon during a lunar eclipse.

In 280 BC Aristarchus of Samos (c. 310– c. 230 BC) proposed the Sun to be the centre of the Universe, with the stars unimaginably distant, but his idea was not generally accepted.

THE EARTH-CENTRED UNIVERSE

PTOLEMY (*c.* AD 120–180) was one of the last great Classical astronomers. He developed a system explaining the complex motions of the planets in the sky. Seen from Earth, the outer planets – Mars, Jupiter and Saturn – occasionally reverse in their path across the sky. In his great book *Almagest* ('Great Synthesis'), Ptolemy suggested they were orbiting in epicycles, moving around a circle, whose centre was itself moving around the Earth. *Almagest* was a masterpiece of complex geometry.

Ptolemy's system of the Universe, with the Earth at its centre, was regarded as unshakeable truth until the sixteenth century and the Copernican Revolution.

Ptolemy's Earth-centered Universe is in direct opposition to Copernicus's heliocentric model.

Controversial Theories

THE COPERNICAN REVOLUTION

THE POLISH astronomer Nicholas Copernicus (1473–1543) challenged the most fundamental basis of astronomy: that the Earth was the absolute centre of the Universe. This view had been accepted for nearly 1,500 years, but Copernicus found that the geometry invoked to explain the movements of the planets was too complex and contrived to be natural. Instead, he tackled the problem mathematically, using his own naked-eye observations of the five planets that were known to exist at that time (Mercury, Venus, Mars, Jupiter and Saturn), to construct a new model of the Solar System. He published his great book, *De Revolutionibus Orbium Coelestium* ('On the Revolutions of the Heavenly Spheres'), in 1543, in which he correctly and firmly placed the Sun at the centre of the Solar System, with the planets orbiting around it. Fearing recriminations by the Church, he delayed

Copernicus's planetary system revolutionised cosmology by placing the Sun at the centre of the Universe.

the publication until the end of his life (legend has it that he finally saw a printed copy while lying on his deathbed).

GALILEO

THE ITALIAN astronomer Galileo Galilei (1564–1642) was a firm supporter of the new Copernican Sun-centred model of the Solar System, which had been slowly gaining support. In 1610, with his new telescope, he discovered that Jupiter had four moons of its own, proving that not everything orbited the Earth. Copernicus's claims were further backed up by Galileo's observations of the phases of Venus in 1610, which were best explained if it revolved around the Sun. Galileo's claims greatly angered the Church, and in 1632 he was forced, under threat of torture, to renounce the Sun-centred Solar System. With his health failing, Galileo was imprisoned in his house by the Church, and died 10 years later. The Catholic Church formally apologised for its treatment of the brilliant astronomer in October 1992. The importance of Galileo's observations and the influence of these on the work of later astronomers and astronomy in general was truly great.

Nicholas Copernicus

The Polish astronomer Nicholas Copernicus (1473–1543) waited until shortly before his death to publish his most famous work. *De Revolutionibus Orbium Coelestium* challenged the Earth's supremacy at the centre of the Universe, and invited the wrath of the Church. His (correct) theory was that the Earth spins on its axis and revolves around the Sun.

The Big Bang

MOST ASTRONOMERS believe that the Universe began in a cataclysmic explosion referred to as the Big Bang. Exactly when this moment of creation occurred is one of the hottest topics in modern astronomy, but estimates range from around 11 billion to 18 billion years ago. All the matter in the Universe began as an infinitely dense, hot, rapidly expanding fireball. As this inferno spread out and cooled, stars, galaxies and eventually planets formed from the original matter.

So where did this explosion occur? Outside the Big Bang was absolutely nothing: length, width and height, as well as the fourth dimension of time did not even exist. Rather, the entire cosmos existed at one point, with all the dimensions beginning within. In 1964 two American scientists, Arno Penzias (b. 1933) and Robert Wilson (b. 1936) discovered that the whole Universe was radiating microwaves. As the Universe expanded to its current conditions, the flash of the original fireball thinned out and cooled, leaving behind these microwaves, which are rather like a ghostly echo.

Arno Penzias and Robert Wilson, who won the Nobel Prize in 1978, pose in front of their radio telescope.

THE FATE OF THE UNIVERSE

THE UNIVERSE may not expand forever. If there is enough matter in the Universe, billions of years from now, its combined gravity may halt the expansion and draw everything back together, destroying everything in what astronomers call the 'Big Crunch'.

The alternative is not much more attractive. The countless stars are gradually consuming the hydrogen in the Universe. Eventually no new stars will be able to form and there will only be the cooling cinders of dead stars, and black holes, in a cold, lifeless Universe, growing larger and ever-more diffuse.

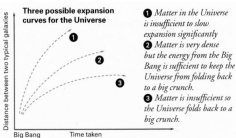

Three possible expansion curves for the Universe

Distance between two typical galaxies

Big Bang   Time taken

❶ *Matter in the Universe is insufficient to slow expansion significantly*
❷ *Matter is very dense but the energy from the Big Bang is sufficient to keep the Universe from folding back to a big crunch.*
❸ *Matter is insufficient so the Universe folds back to a big crunch.*

Fred Hoyle

Fred Hoyle (b. 1915) was one of the key developers of the Steady State theory of the Universe, although most astronomers now believe in the Big Bang. The Steady State theory said that the Universe was ever expanding, with no beginning or end, and with new matter appearing to fill the gaps left by the expansion.

LOOKING AT SPACE

Introduction

LONG BEFORE the invention of the telescope in the early seventeenth century, civilisations such as the Mayans of South America built observatories from which to chart the motions of celestial bodies. The Mayan civilisation was at its height around 300 BC and they developed calendars with which to predict seasonal weather by closely studying the changing movement of the Sun in the sky over the course of a year.

In England, Salisbury Plain saw the great stone circle of Stonehenge built in stages between 3100 and 1500 BC and, although its exact purpose is not known, it was built to precise astronomical alignments. Some of the stones appear to have been placed to show the positions of the Sun and Moon when rising and setting. It is possible that these arrangements allowed the community there to predict the longest and shortest days of the year, and even eclipses.

An ancient place of Celtic worship, Stonehenge also appears to have functioned as a sophisticated celestial observatory.

THE TELESCOPE

THE ASTRONOMICAL TELESCOPE was one of the great instruments of the scientific revolution of the seventeenth century. Glass had been discovered by the Phoenicians around 3500 BC, but it was another 5,000 years before a lens good enough for a telescope was made. A Dutch spectacle maker, Hans Lippershey (c. 1570–1619), built a telescope in 1608 (it is not known for certain if his was the first). Word of the new device raced across Europe and inspired Galileo Galilei to build his own in 1609. His observations of the phases of Venus and the moons of Jupiter were instrumental in dethroning the Earth as the centre of the Universe. Telescopes developed rapidly and in 1668 Isaac Newton (1642–1727) constructed the first reflecting telescope – the ancestor of today's modern observatories.

Galileo Galilei

Galileo Galilei (1564–1642), born near Pisa, Italy, was the first astronomer to use the telescope systematically. He invented a telescope which enabled him to discover the moons of Jupiter, the phases of Venus and the mountainous nature of the surface of the Moon. His discoveries infuriated the Church, which placed him under house arrest. The Vatican only admitted its mistake a few years ago.

Optical Telescopes

TELESCOPES ARE DEVICES that make distant objects appear much larger and brighter than normal. The first and most common type of telescope is the optical telescope, which deals purely with the wavelengths of light visible to the human eye. The earliest telescopes were called refractors and these worked using lenses made from glass. The main lens, which captures the starlight, is the objective lens. This focuses the light on to an eyepiece at the other end of the telescope tube. The larger the objective lens, the higher magnification eyepiece can be used while keeping a good quality image.

The 71-cm (28-in) refractor telescope at the Royal Observatory in Greenwich.

A more complex type of telescope is the reflecting telescope. Instead of a large objective lens, this type of instrument uses a concave mirror to focus light from the object in view on to the magnifying eyepiece. One of the advantages of reflecting telescopes over refractors is that the length of the telescope tube does not have to be very long, and therefore it is less cumbersome to use and more portable.

The most important attribute of any telescope is

its light-gathering power. In order to use a powerful eyepiece to obtain a very high magnification, the telescope has to gather as much light as possible, or the resulting image will be very dim. The more powerful telescopes tend to be reflectors, as it is easier to construct a large mirror than a large lens. Despite this, some of the greatest astronomical discoveries have been made using giant refractors. The Greenwich Royal Observatory has one of the most famous. Its lens is 71-cm (28-in) across, and it is the seventh-largest refractor in the world. Since it first opened in 1894, it has made significant contributions to our knowledge of binary stars, although its observations have been at a more basic level for the past 50 years.

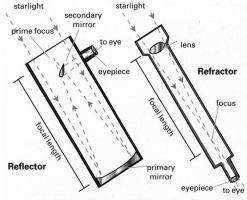

The difference between a reflector and a refractor

Optical Telescopes –
the Bigger the Better

THE OPTICAL TELESCOPE, from its humble beginnings in the early seventeenth century, has come a very long way. The last few decades have regularly seen the world's best telescopes being outclassed by better instruments. The most powerful of today have light-gathering mirrors of previously undreamed-of proportions.

The Keck Observatory telescope is the most powerful currently in operation. Situated at an altitude of 4,200 m (13,800 ft) on the summit of Mauna Kea, Hawaii, it has access to some of the clearest skies above Earth. Keck consists of two telescopes, each with a mirror 9.82 m (32.2 ft) across. The mirrors consist of 36 separate segments, each of which can be moved to an accuracy of 1/1000th of the width of a human hair by pistons underneath. The ability of the pistons to move prevents the mirrors from deforming under their own weight.

One of the biggest telescopes in the world inside the Keck Observatory, Hawaii.

Although these two giant telescopes operate independently, the next step will be for them to combine the starlight they gather using a technique called interferometry, allowing them to behave as if they were effectively one giant telescope.

THE VERY LARGE TELESCOPE

THE TWIN KECK telescopes are not the only large optical interferometer planned. On a lonely mountain top in Chile, the imaginatively named Very Large Telescope (VLT) consists of four mammoth telescopes, each with a mirror 8.2 m (27 ft) across. Work has now started on three smaller auxiliary telescopes, which will work alongside the four huge reflectors. In a few years time, when the VLT is complete, all seven telescopes will operate as an interferometer, combining their starlight. If all goes well, this optical monster will have the effective resolution of a mirror 130 m (426.5 ft) across.

Searching deep into the skies: the VLT, Chile.

Hubble Space Telescope

NAMED AFTER the American astronomer Edwin Hubble (1889–1953), the Hubble Space Telescope is the most powerful instrument we have ever had to scrutinise the heavens. In an orbit 600 km (373 miles) above the Earth it is high above our turbulent atmosphere, which inevitably blurs the views of ground-based optical observatories.

Soon after its launch in April 1990, it became apparent that Hubble was suffering from a condition known as spherical aberration – an inability of the 2.4 m- (7.8 ft-) wide primary mirror to focus the light it gathered properly. This simple and highly embarrassing design flaw had to wait more than three years to be corrected. Fortunately, the National Aeronautics and Space Administration (NASA) intended Hubble to be visited every few years by astronauts to be upgraded. In December 1993, the crew of the space shuttle Endeavour fixed the problem and allowed Hubble to view our Universe with unprecedented clarity. Since then, the images taken by this great space observatory have stunned both the public and astronomers alike.

The Hubble Space Telescope (HST), deployed on 24 April 1990, orbiting around Earth.

SEEING FURTHER

THE MOST SPECTACULAR results have been obtained
with the on-board Wide Field and Planetary Camera and the
Faint Object Camera, revealing details 10 times finer than
conventional ground-based telescopes. Hubble has peered into

the heart of the Orion
Nebula and found dusty
disks surrounding forming
stars – a sign that many, if not
most stars are probably
accompanied by planets.
Hubble has imaged
unmistakable signs of super-
massive black holes lurking at
the hearts of other galaxies.
In our own Solar System,
Hubble has monitored dust
storms on Mars, and has
found many new small icy
bodies at the very edge of
our Solar System.

*A composite image from HST of the
gas dust pillars in the Eagle Nebula; the
black areas are beyond the cameras'
reach of information.*

Edwin Hubble

US astronomer Edwin Hubble (1889–1953) is best
remembered for discovering the relationship between the
distances of galaxies and their speed, as they move away
from us in the expanding Universe. One of the main goals
of his namesake, the Hubble Space Telescope, is to further
study this relationship – known as the Hubble Constant.

Beyond the Human Eye

T HE VISIBLE LIGHT we can see is just a tiny fraction
of the whole electromagnetic spectrum. Until recently, all our
astronomical knowledge was based on visible light. But today,
astronomers use the whole spectrum to learn about the Universe.

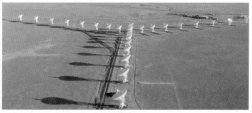

*An aerial view of the 27 satellites dishes making up the Very Large Array
in New Mexico, USA.*

RADIO ASTRONOMY

RADIO WAVES have much longer wavelengths than visible
light, and can penetrate through the dust clouds in the Galaxy.
In 1932, using long radio antennae, the engineer Karl Jansky
(1905–50) first discovered natural radio waves coming from the
centre of the Galaxy, in the constellation of Sagittarius. Since
then, bigger and better radio telescopes have been constructed
to map the Universe at radio frequencies. Modern radio
telescopes use a large parabolic dish to focus the gathered radio
waves on to a detector, and many radio dishes can be linked
together to act as one interferometer – just like the optical
telescopes of the VLT in Chile.

MICROWAVE ASTRONOMY

NASA LAUNCHED the Cosmic Background Explorer (COBE) satellite in 1989. Its goal was to improve our understanding of cosmology, the study of the beginning, evolution, and structure of the Universe as a whole. Over five years COBE studied the very faint background microwave radiation, which comes from all directions in the Universe. This background radiation is an echo of the colossal explosion at the very start of the Universe, the Big Bang. The colour-coded map produced by COBE shows minute temperature variations in this low-temperature radiation, suggesting that the very early Universe was not completely smooth, but had regions slightly more dense than the rest. These regions are likely to have been the 'seeds' of the billions of galaxies in the Universe today.

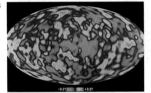

A COBE map of cosmic background radiation, showing the temperature variations that have led scientists to conclude that the early Universe had areas of different densities.

Karl Jansky

Karl Jansky (1905–50) was one of the pioneers of radio astronomy. He joined the Bell Laboratories, New Jersey, in 1928, where he was assigned the task of analysing the static which accompanied long-distance telephone calls. This work led to his discovery of natural radio waves coming from the centre of the Galaxy.

Beyond the Human Eye – Going Further

ULTRAVIOLET ASTRONOMY

STARS EMIT MUCH of their energy as ultraviolet radiation, which exists just beyond the violet part of the spectrum that we can see. The Earth's ozone layer filters out most of the ultraviolet rays from space, so astronomers have used orbiting satellites to study them. Ultraviolet astronomy has been crucial to our understanding of the Sun and planets, as well as hot stars and the energetic centres of galaxies.

X-RAY ASTRONOMY

IN JULY 1999 NASA launched the Chandra X-Ray Observatory. High above our atmosphere, which absorbs X-rays, Chandra analyses the X-radiation from gases in the Universe with temperatures of millions of degrees, and converts it into a picture that we can see.

Chandra has provided spectacular new images of the invisible Universe, including extremely hot gases spiralling around the tiny pulsar at the heart of the Crab Nebula.

GAMMA-RAY ASTRONOMY

THE STUDY OF gamma rays began in 1967 when they were discovered accidentally by the American Vela satellites, which were monitoring Earth for signs of nuclear explosions. Gamma rays are the most energetic form of radiation and, like X-rays, can only be studied in detail from space. In 1991 NASA launched the Compton Gamma-Ray Observatory into space. Since its launch

t has provided new insights into the most violent processes in the
Universe. Of particular importance were the new views of
'gamma-ray bursts'. Lasting between a few milliseconds and a few
minutes, the immense power of these events is second only to the
Big Bang, and they may signal the formation of black holes in
distant galaxies, billions of light years away.

The Crab Nebula is the remnant of a supernova, as observed by Chandra.

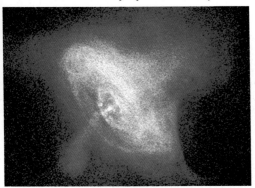

Jocelyn Bell Burnell

Jocelyn Bell was born in Belfast in 1943. Her father was
the architect for the Armagh Observatory and her interest in
astronomy was spurred on by staff at the observatory. She
became famous for the first discovery of a pulsar, the
supernova remnant at the heart of the Crab Nebula, in 1967.

Capturing Images of Space

ASTROPHOTOGRAPHY

FOR OVER 200 YEARS after the invention of the telescope, astronomers recorded their observations by sketching what they could see. When the camera was developed in the nineteenth century, it caused a revolution in astronomy. In around 1840 a specially constructed camera was attached to a telescope for the first time. The first astronomical photographs were crude, but techniques developed rapidly. Photographs provided permanent, accurate records of the sky and today millions of people can enjoy the beauty of the Universe without the need of a telescope of their own.

Most astronomical objects are incredibly faint, but a camera can be pointed at a faint gas cloud and its shutter left open for long periods of time. In these long exposures, the camera film can slowly accumulate the faint light from an object and record fine details invisible to the naked eye.

The Pleides star cluster, containing several hundred stars, is a conspicuous object in the night sky.

DIGITAL IMAGING

PHOTOGRAPHIC FILM, however, only registers around one per cent of the light that hits it. For the last few decades, new, more efficient electronic cameras called charge-coupled devices (CCDs) have been used on the world's largest telescopes, as well as by amateur astronomers. In a CCD, the light detector is a grid-like array of tiny electrodes. When starlight hits these electrodes, a small electric charge collects behind each one, proportional to the amount of light that it receives. Once an observation is complete, all the electric charges for each electrode are measured and converted into a digital version of a photograph. This image can then be processed, using a computer, to enhance contrast, remove defects and add false colours to highlight particular parts of the image.

An original image of the comet Hale-Bopp (left) is enhanced through a filter (right) to show varying thermal densities.

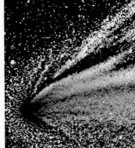

MEASURING SPACE

Introduction

THE SPEED OF LIGHT

FROM ERATOSTHENES' (276–194 BC) first accurate measurement of the circumference of the Earth, the measurement of space has been the backbone of astronomy. In 1675 the Danish astronomer Ole Römer (1644–1710) attempted to measure the speed of light. By timing the moons of Jupiter being eclipsed by the shadow of the planet, he discovered that the prediction charts of these eclipses were inaccurate. He realised that this was due to the amount of time it took for the light from these events to reach Earth. His calculation of the speed of light was three-quarters that of the actual value of 300,000 km (186,420 miles) per second.

Tycho Brahe

The Danish astronomer Tycho Brahe (1546–1601) made the best observations of the heavens before the invention of the telescope. He spent 20 years meticulously plotting the motions of the planets. He never accepted that the Earth orbited the Sun, and proposed that the other planets circled the Sun, which in turn orbited Earth.

THE SCALE OF SPACE

FROM OUR VANTAGE-POINT on our tiny Earth, the Universe is unimaginably vast. Because of the finite speed of light, we see astronomical objects as they were in the past. We see the Moon as it was one-and-a-quarter seconds ago, and the Sun, 149,600,000 km (92,961,440 miles) away, as it was around eight minutes ago. On average, Jupiter is 43 'light minutes' away and Pluto is some five-and-a-half 'light hours' away. Beyond the Solar System, the scale increases dramatically. It would take 4.25 years to reach the nearest star to the Sun – Proxima Centauri – at the speed of light. Normal units of measurement become cumbersome at this point, so instead of using kilometres or miles, stellar distance (distances between the stars) is measured in light years. To cross the Milky Way Galaxy would take over 100,000 years, and our nearest other major galaxy, Andromeda, is over two million light years away. To travel to the edge of the observable Universe at the speed of light could take a staggering 12 thousand million years.

A Hubble deep-field image of directly observed galaxies that are some eight billion light years away.

Early Maps of the Sky

THE PREHISTORIC SKY

IN 2000, ARCHAEOLOGISTS uncovered what appears to be a crude ancient map of the sky in the Lascaux caves in France, famous for their prehistoric drawings. Believed to be 16,500 years old, this cave painting contains what looks like early representations of the constellations Lyra, Cygnus and Aquila. There is also a painting of a cluster of stars above a bull, strikingly similar to the Pleiades star cluster in the constellation of Taurus, the bull. This discovery reveals the significance with which prehistoric people regarded the sky.

Our ancient ancestors also made maps of the Moon. In 1999 a series of shapes was discovered, carved into a rock at Knowth, one of the most ancient Neolithic sites in Ireland. Superimposed over a picture of the Moon, the stone markings line up with the dark lunar seas. At around 5,000 years old, this map is 10 times older than Leonardo da Vinci's (1452–1519) sketch from 1505, previously regarded as the oldest Moon map.

THE CONSTELLATIONS

GREAT ANCIENT CIVILISATIONS such as the Chinese, Egyptian and Chaldean divided the sky into constellations (groups of stars), and gave them mythical names, depending on their shapes in the sky. The constellations were formalised by the Greek astronomer Ptolemy, whose 48 shapes and names we still use today, along with 40 others from more modern times.

The constellation Taurus in the sky: the dotted position of the stars connect to form the image of the bull.

Star maps as scientific tools did not come about until the second century BC. The Greek astronomer Hipparchus (190–120 BC) witnessed a temporary star in the sky and made the first accurate chart of the heavens, in order to be able to detect any similar changes.

Star Catalogues

WHEN HIPPARCHUS made his chart of the sky in the second century BC, he catalogued the positions and relative brightness of 1,025 stars. This meticulous feat of observation led to the discovery of the precession of the equinoxes – the slow change in the orientation of the Earth as it wobbles like a spinning top around its axis of rotation.

John Flamsteed (1646–1719) was appointed Britain's first Astronomer Royal in 1675. He compiled a new catalogue of the stars, for both navigation and astronomy. It was published in 1725, five years after his death, and contained accurate information on 3,000 stars. In 1828 Caroline Herschel (1750–1848) received the gold medal of the Royal Astronomical Society for compiling and editing Flamsteed's catalogue, along with the observations of her brother, William Herschel (1738–1822). French astronomer Charles Messier (1730–1817), while studying 19 comets he had discovered, noticed

A telescopic photograph of the Crab Nebula (M1), the gaseous supernova remains of an ancient star, in Taurus.

a patch of light in Taurus. This object became known as Messier 1 (M1), the Crab Nebula, the debris from a supernova that had exploded in AD 1054. His catalogue of nebulae and star clusters was published in 1781, the numbers from which are still in use today.

A MODERN VIEW

THE MOST UP-TO-DATE star catalogue has been compiled by the European Hipparcos satellite, which operated from 1989 to 1993. Hipparcos was capable of measuring the height of an astronaut standing on the Moon, as seen from Earth. During its mission it plotted the positions and brightnesses of over 100,000 stars with pinpoint accuracy, and many more stars with less detail but still more precisely than ever before. As the stars slowly orbited the centre of the galaxy, Hipparcos was able to monitor the tiny relative motions of the stars across the sky. This movement, called 'proper motion', is invisible to the naked eye but causes the constellations to slowly change in shape over thousands of years.

John Flamsteed

John Flamsteed (1646–1719) was England's first Astronomer Royal, appointed in 1675 by Charles II to run the newly built Greenwich Royal Observatory. He spent over 40 years compiling the first comprehensive catalogue of the stars and observations of the movement of the Moon, both of which were essential to improving sea navigation.

Models of Space

EUDOXUS OF CNIDUS

OFTEN DESCRIBED as one of the fathers of modern astronomy, Eudoxus of Cnidus (408–355 BC) was the first ancient astronomer to make a map of the stars. Born in the last years of the fifth century BC, he founded a school of astronomy in Cyzicus, Asia Minor, and later studied at the academy of Plato.

He is best known for proposing a heliocentric (Sun-centred) model for the Solar System, with the six known planets being supported by concentric spheres around the Sun. He was the first astronomer who truly attempted to explain the motions of the planets using scientific methodology.

ARMILLARY SPHERES

ARMILLARY SPHERES were the first devices used to represent the heavens realistically. They depict the whole sky as a concentric shell around the Earth. The concept of the celestial sphere was crucial to ancient astronomy, and is still a fundamental scientific tool today for mapping the heavens using the celestial equivalents of latitude and longitude.

This seventeenth-century armillary sphere is a crude model of the motions of the Sun, stars and planets.

There are two different types: one used for observation and a more complex version for demonstration purposes. The first demonstration model may have been built by Archimedes (287–202 BC) and the first reference to an observational sphere was made by Ptolemy (*c.* AD 120–180).

Orreries were early mechanical models of our Solar System that simulated the movement of the planets around the Sun.

ORRERY

THE FIRST ORRERY was made in the eighteenth century. They were named after the fourth Earl of Orrery who was a patron of their inventor, George Graham (*c.* 1674–1751). These beautiful models show the positions of the planets and their major moons, with the Sun at the centre. Powered by a clockwork mechanism, the model planets revolve around the Sun at the correct relative speeds – the orrery was essentially the first ever planetarium.

Friedrich Bessel

THE GERMAN ASTRONOMER and mathematician Friedrich Bessel (1784–1846) is best remembered for calculating the first reasonably accurate distance to another star. He used a technique known as parallax – a term that is familiar to surveyors. If you hold up your finger at arm's-length and look at it with one eye and then the other eye, it appears to move relative to background objects. If your finger is closer, the apparent shift is greater. Provided you know the distance between your eyes (the baseline), and can measure the shift, simple trigonometry can be used to calculate the distance to your finger.

THE FIRST CELESTIAL DISTANCE

BESSEL EXPLOITED this principle on a massive scale. He chose the star 61 Cygni, which he believed to be a relatively nearby star, and for his baseline he used the diameter of the Earth's orbit around the Sun. He measured precisely the position of 61 Cygni compared to the background stars, and again six months later, once the Earth had completed half of one orbit around the Sun. Just as he expected, this star appeared to move slightly back and forth when viewed from opposite sides of the Earth's orbit – a baseline distance of almost 300 million km (187 million miles).

The star 61 Cygni had moved only a tiny distance in the sky, just 0.29 seconds of arc (a unit of measurement on the sky), and Bessel calculated that this corresponded to a distance between Earth and the star 61 Cygni of 11.2 light years. Although this method works only for the nearest stars, Bessel's announcement of his work in 1838 caused a revolution in astronomy. For the very first time, real distances to objects outside our Solar System could be measured, and the true scale of the Universe began to dawn on the world.

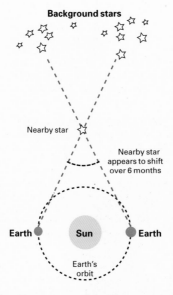

Background stars

Nearby star

Nearby star appears to shift over 6 months

Earth

Sun

Earth

Earth's orbit

Parallax is used by astronomers to calculate distances to stars.

The Expanding Universe

OTHER GALAXIES

WHEN THE third earl of Rosse (1800–67) used his newly completed 1.8-m (72-in) telescope to observe a nebula in the constellation Canes Venatici in 1845, he noticed it had a spiral shape. Many more of these strange objects were soon discovered and astronomers were puzzled about their nature.

In 1923, Edwin Hubble analysed the light from these nebulae and discovered that they were made of hundreds of billions of individual stars. From the brightness of these stars, Hubble calculated that they were millions of light years away and that these spirals were independent galaxies, like our own Milky Way.

THE FATE OF THE UNIVERSE

FURTHER ANALYSIS of these galaxies revealed that the light in their spectra were redshifted due to the Doppler effect. We experience the Doppler effect every day, only with sound instead of light. The pitch of a siren is higher when an ambulance is heading towards us and lower once it has passed. The redshifted light from the galaxies means they are all moving away from us, and away from each other.

A breathtaking image of M51, the Whirlpool galaxy, captured by the Hubble Space Telescope.

Furthermore, the speed of a receding galaxy is directly proportional to its distance. The logic was inescapable: the Universe, made up of countless galaxies, is expanding. Astronomers now believe that the whole Universe began in a colossal explosion, around 11–18 billion years ago. We call this event the Big Bang, and the Universe has been expanding ever since. Finding the exact rate of expansion is one of the greatest quests in modern astronomy, and the answer will tell us if the Universe will continue forever, or stop expanding as its gravity causes it to collapse and self-destruct in a 'Big Crunch'.

If you have a source of light of a known colour, the Doppler effect can be used to determine its velocity in respect to you by the colour you observe. If the light you see is redder than you know the original colour to be, then the wavelengths are longer and the light is moving away from you (A); if the light is bluer, then the wavelengths are shorter, and it is moving towards you (B). Here, the source is moving to the left.

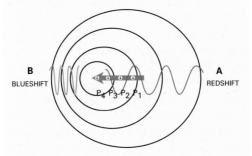

SPACE AND OUR WORLD

Introduction

THE EFFECTS of the heavens on our world have always been present in the history of civilisation. Prehistoric man used the positions of the rising Sun and stars to judge the changing seasons, allowing them to be efficient farmers. But our ancestors also believed that the night sky had a more direct impact on our daily lives. During the Middle Ages, the study of

During the Middle Ages, the belief in the power of the heavens to predict people's destinies was unshakeable, represented here in the signs of the zodiac.

astronomy and astrology were one and the same, and astrologers used the night sky to make predictions about the future. The Sun and planets were believed to be gods, and their relative positions at the time of someone's birth were thought to have major influences on that person's life. In fact the word 'disaster' comes from the Latin words for 'bad star', often referring to the appearance of comets heralding bad luck.

Today we no longer worship the Sun and planets, and the study of astrology is considered by most people to be merely a superstition, carried on despite the advances of modern science.

SPACE WEATHER

SPACE, HOWEVER, does directly affect our daily lives. For example, the Sun will occasionally spit out huge eruptions of ionised gas or plasma in vast clouds, called coronal mass ejections (CMEs). When these huge clouds reach Earth, they cause vivid auroral displays (the northern and southern lights), as the solar particles slam into our upper atmosphere and cause it to glow like a neon tube. With modern society relying increasingly on technology, this 'space weather' represents a serious threat: the electronics of orbiting satellites can be disabled by these eruptions, and whole cities could be blacked out due to electro-magnetic interference.

The Aurora Borealis phenomenon (the Northern Lights): curtains of light caused by particles of ionized gas from the Sun hitting the Earth's atmosphere.

Newton's Theory of Gravity

IT WAS JOHANNES KEPLER (1571–1630) who first attempted to find simple geometric rules that governed the motions of the planets. He began his work in 1596, and in 1609 and 1619 he published his fundamental laws. His three laws described the orbits of the planets as being elliptical, and gave the direct relationship between the speed of their orbits and their distances from the Sun. Kepler's laws gave the first reasonably accurate scale to the Solar System, but an explanation of why they worked had to wait until Isaac Newton published his ground-breaking book *Principia Mathematica* in 1687.

According to history, Isaac Newton was inspired by seeing an apple fall from a tree in his garden, and he realised that the force which kept the planets moving around the Sun was the same one that made objects, like the apple, fall to the ground.

Newton discovered that each object in the Universe is attracted to every other one by a force called gravity. The strength of this attraction between bodies depends on their distances away from each other and how massive they are. A good example is the Earth

Newton's famous book, Principia Mathematica, *explained the mechanics of the Solar System in mathematical terms.*

and the Moon: if you doubled the Earth's mass then the Moon would be pulled twice as strongly by Earth's gravity. But if you double the distance between them, then the Moon is pulled only one quarter as strongly.

Newton had finally provided the physics behind Kepler's laws and the movements of the planets, and proved that it was possible for us to use science to decode and understand the laws which Nature herself follows.

Isaac Newton

Isaac Newton (1642–1727) was born in Lincolnshire, UK, in the year that Galileo Galilei died. He entered Cambridge University in 1661 and built the first-ever reflecting telescope in 1668. At the age of just 27, Newton became professor of mathematics at Cambridge. As well as his theory of gravity, Newton made breakthroughs in understanding the nature of light.

Johannes Kepler

German astronomer Johannes Kepler (1571–1630) demonstrated that the planets orbit the Sun in an ellipse, not a perfect circular path. Kepler studied under the Danish astronomer Tycho Brahe and succeeded him in 1601 as Imperial Mathematician to Emperor Rudolf II in Prague. Kepler's third law of planetary motion formed the basis of Isaac Newton's law of gravity.

Latitude and Longitude

ANCIENT SAILORS depended on landmarks to navigate and never ventured far from the coastline. But at the end of the fifteenth century, Portuguese sailors began to navigate using the stars. Polaris, the pole star, lies directly above the North Pole, along the axis of the Earth's rotation. As the Earth rotates, the sky traces a circle around the stationary Polaris. By measuring the angle of Polaris above the horizon, using a sighting device called an astrolabe, sailors could calculate their north–south position on the Earth – i.e., their latitude.

THE LONGITUDE PROBLEM

IN ORDER to be able to navigate accurately, sailors also needed to know their east–west location, or their longitude. As ships travelled across the globe, the local time moved back one hour for every 15 degrees they sailed west, and one hour forward when travelling east (the circumference of the Earth is 360 degrees). This meant navigators potentially could work out their longitude by observing the Sun, and

The mariner's astrolabe, a naked-eye instrument, was used to measure the positions of stars.

omparing the local time to the time at another reference point
n Earth, like Greenwich. They kept Greenwich time on board,
out the timepieces of the early eighteenth century swiftly lost
accuracy as the ships tossed on the ocean.

Spurred on by a prize of £20,000, the Yorkshireman John
Harrison (1693–1776) began building clocks designed to be
immune to the motion of a ship. He built his first prototype
between 1730 and 1735, and in 1762 Harrison tested his latest
device, the 'H4', on a voyage to Jamaica. It kept accuracy to
within 5.1 seconds. The longitude problem had been solved,
but it was not until 1773 that Harrison received the full prize
money from the British government.

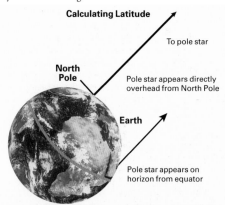

Calculating Latitude

To pole star

North Pole

Pole star appears directly
overhead from North Pole

Earth

Pole star appears on
horizon from equator

*The pole star, Polaris, was used by sailors together with the astrolabe to
calculate their latitude.*

Using Satellites to Map Earth

SINCE THE DEVELOPMENT of space rockets and satellites, one of the most significant uses of space has been as a high vantage point from which to survey the Earth. The most successful programme of satellite surveys has been Landsat. Since 1972, seven Landsat satellites have provided images of the Earth from space, showing details around 33 m (108 ft) across on the ground. As well as visible images, Landsat also surveys the Earth's surface in infrared light. The latest model, Landsat 7, was launched in April 1999 and can see details as small as 15 m (50 ft) across.

The Landsat series has acquired millions of images of the surface of Earth from space. This priceless resource has played a crucial role in monitoring environmental disasters. Countries such as Australia and Pakistan have used Landsat data to reveal the extent of major floods, and the infrared images of forests in the USA have allowed fire fighters to assess the risk of fires, by detecting especially dry and flammable areas. In 1975, a US research ship was guided through the Antarctic pack ice using channels of open water revealed in Landsat images.

The French space agency, Centre National d'Etudes Spatiales (CNES), also has a history of successful Earth resource satellites. The Systeme Pour l'Observation de la Terre (SPOT) programme consisted of four satellites, the first of which, SPOT 1, was launched in 1986. This series of remote-sensing satellites has been used in a variety of Earth observation programmes, monitoring pollution, erosion and deforestation, with sensors able to see some surface features 10 m (33 ft) across.

A satellite image of the floods in Mozambique, 2000.

Space Technology on Earth

CONTRARY TO modern mythology, neither teflon nor velcro are spin-offs from space technology, and were not invented by NASA! However, the number of everyday objects and technologies that have been developed or enhanced as a direct result of space exploration is astonishing. According to NASA, more than 30,000 space-derived innovations affect our daily lives.

EVERYDAY ITEMS

THE SCRATCH-PROOF LENSES of sunglasses have been adapted from the protective layer which coat the visors of astronauts' space helmet visors. Also, the waterproof material Gore-Tex, used widely in outdoor pursuits, was initially developed for spacesuits. Many DIY practitioners owe some of their favourite tools to space exploration: the first cordless drill was built for use in obtaining core samples on the Moon.

HIGH-TECH ADVANCEMENTS

IN MEDICINE, techniques such as laser surgery and brain scans have been enhanced from space research. The equipment that allows doctors on Earth to monitor astronauts' life-signs has been applied to life-support machines in intensive care units. More recently, the

The advances made in computer technology for the Apollo flights (pictured) had a direct influence on the hi-tech equipment that we now take for granted.

digital imaging used by the Hubble Space Telescope has led to new non-invasive surgical techniques to be used by doctors in the fight against breast cancer.

A new car has far more computing capability than an Apollo spacecraft. We owe much of our modern computers to the giant leaps made in the 1960s, when even a primitive processor had to be small and light enough to fit inside the lunar landers. It is difficult to compare the value of all the benefits we gain from space-derived technology, but it seems that we are unable to use a computer, make a phone call, or visit the doctor without running into a descendent of space exploration.

Lunar rovers such as this one used revolutionary technology.

Eclipses

TOTAL ECLIPSE OF THE SUN

BY A CURIOUS COINCIDENCE of Nature, the Sun and the Moon appear the same size in our sky. While 400 times the width of the Moon, the Sun is also 400 times further away. When the Moon passes directly between the Earth and the

Sun, a few times each year, its shadow races across the Earth at over 2,000 kph (1,243 mph). Anyone in its path (which can be up to 250 km/155 miles wide, (although the average width is more like 160 km/100 miles) will witness one of the most spectacular natural phenomena – a total solar eclipse. The Moon slowly covers up the Sun until, 10 minutes before totality, the light and temperature drop noticeably. The Sun is now a slim crescent and as the last rays of the Sun shine through the valleys on the Moon's jagged edge, they form a diamond-necklace effect. At totality the Sun disappears and its ghostly outer atmosphere – the corona – becomes visible, stretching into space. Within a few minutes, the Moon's shadow passes, and totality ends. The Moon's orbit is elliptical, and if an eclipse occurs when the Moon is at its furthest distance from the Earth, it cannot cover the entire Sun, and so we see an annular eclipse. This is when the Moon passes directly between the Sun and the Earth, and appears as a dark disc surrounded by a narrow ring of sunlight.

LUNAR ECLIPSES

MUCH LESS DRAMATIC, but still beautiful, are eclipses of the Moon, when the Earth passes directly between the Moon and the Sun.

The curved shadow of the Earth begins to creep across the lunar surface until the whole Moon is in shadow. During totality, which can last up to 1 hour 47 minutes, the Moon does not actually disappear, but is dimly lit by sunlight that has been refracted on to it by the Earth's atmosphere, bathing the Moon in a magnificent coppery light.

A total solar eclipse: the Moon is covering the entire disc of the Sun.

Meteorites

IN JULY 1994, people around the world watched as the fragments of comet Shoemaker-Levy 9 impacted with Jupiter in a series of spectacular explosions. This celestial event reminded us that we too are vulnerable to impacts from space. One look at the Moon shows that it has been struck countless times in its history. The same is true for our planet, but the weathering effects of the atmosphere, coupled with

A Hubble image showing the aftermath of comet Shoemaker-Levy 9's impact on the atmosphere of Jupiter.

Earth's geological activity, has erased the evidence of bombardment in the distant past. Most comets and asteroids were swept up by the planets during their formation 4.6 billion years ago but, as Shoemaker-Levy 9 proved, there are still plenty of objects out there which can collide with the planets. Most of these chunks are tiny and many fall to Earth every year as small meteorites. But occasionally, much larger fragments hit the Earth with devastating results.

IMPACT CRATERS ON EARTH

GEOLOGISTS HAVE catalogued around 150 impact sites on Earth, the most famous of which is the 1.2 km- (¾ mile-) wide Meteor Crater in Arizona, USA. This hole was formed

52,000 years ago, when an iron meteorite only 50 m (164 ft) across hit the ground and exploded with the power of 20 million tons of TNT.

A much more dramatic impact occurred 65 million years ago in the Gulf of Mexico, creating the 180 km- (112 mile-) wide Chicxulub Crater. The global fallout and environmental change from this impact are believed by many to have contributed to the extinction of the dinosaurs.

In 1908, 2,000 sq km (773 sq miles) of the Tunguska region of Siberia were devastated when a comet less than 100 m (328 ft) across exploded in the atmosphere.

Astronomers are gradually discovering and monitoring the thousands of asteroids whose paths cross the Earth's orbit, with the hope of giving advance warning if one is on a collision course.

Meteor Crater in Arizona was formed when a meteorite hurtling through space at 64,000 kph (40,000 mph) struck the rocky plain.

DISCOVERING SPACE

Introduction

EQUIPPED WITH NEWTON'S LAWS of gravity, astronomers can predict the positions of the planets with pinpoint accuracy. As the planets pass close to each other during their orbits, they exert a gravitational tug that slightly disturbs their positions. Such an interaction was noticed to be affecting the orbital position of Uranus after its discovery in 1781, and a new planet, Neptune, was found to be the reason when it was discovered in 1846.

OTHER WORLDS

KNOWING THE MOTIONS of the planets with such precision has allowed small, unmanned space probes to be launched towards them, often taking years to reach their targets. The same gravitational pertubations that led astronomers to Neptune allow probes to adjust and accelerate their journeys across the interplanetary gulfs. These hardy

A close-up of the crater-free surface of Europa, one of Jupiter's moons, as imaged by the Galileo spacecraft probe.

robots have now visited every planet except distant Pluto and mapped alien landscapes, the likes of which had never been imagined. Dried-up river valleys on Mars prove that our planet is not the only one to have had liquid water on its surface. And the icy areas discovered on Jupiter's moon Europa hints at a possible ocean of liquid water under the ice – a potential harbour for life.

MORE QUESTIONS THAN ANSWERS

ON THE GROUND and in space, astronomers have struggled to see ever further into the Universe, with telescopes capable of gauging the full range of natural cosmic radiation, hidden to our limited eyes.

With each leap in telescope power has come new and bizarre discoveries, but there are always more tantalising details, waiting just beyond their limits of telescopic resolution.

The birthplace of stars: a massive dust cloud inside the Eagle Nebula.

Earth and the Solar System

BUILDING THE PLANETS

THE EARTH is just one of nine planets in the Solar System, all of which are very different.

When the planets were forming, over 4.6 billion years ago, their building blocks were of different compositions, depending upon their distance from the Sun. In the inner Solar System, Mercury, Venus, Earth and Mars all formed from dust grains and grew into rocky planets. Further from the Sun, temperatures were cool enough to allow water and other volatile gases to condense into ice crystals. The gas giants, Jupiter, Saturn, Uranus and Neptune, were able to sweep up these ices, as well as the dust, and grow to enormous proportions. While the planets were forming, heavier materials such as iron sank to the centre, forming planetary cores, while lighter substances like water and hydrogen, rose to the surface.

OUR PLANET

EARTH IS A ROCKY planet with a solid iron core, a molten outer core and a rocky mantle, covered by a thin crust of solid rock. Heat from the core leads to extensive volcanic activity on the surface, and plate tectonics make the continents slowly drift across the globe over millions of years.

Earth – the only planet in our Solar System with the right conditions to support life.

Our planet formed in a region of the Solar System that receives just the right amount of heat from the Sun. The inferno of Venus and the frozen wastes of Mars show what happens to worlds formed outside this narrow zone of habitability. The Earth is the only world we know of that has liquid oceans and life, and it provides an excellent comparison for our study of the other bodies of the Solar System.

The nine planets that make up our Solar System.

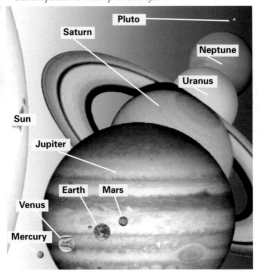

The Sun

OUR SUN IS A STAR around which the planets revolve. Many of the historical milestones in astronomy have been achieved by studying the Sun, from the first reports of sunspots

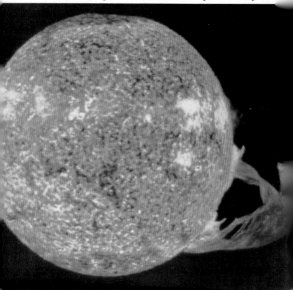

A NASA image of the Sun with a spectacular flare.

by the Chinese in 800 BC to fine details on its surface seen by spacecraft. The Sun is a seething ball of gas, mostly hydrogen, almost 1,400,000 km (870,000 miles) across. The bright gaseous, visible surface – the photosphere – glows at a temperature of 5,500°C (9,900°F). The energy which drives the Sun originates in its core. Here, at its heart, the pressure is around 340 billion times sea-level pressure on Earth, and the temperature is 15,000,000°C (27,000,000°F). Under these conditions, 700 million tonnes of hydrogen atoms fuse into helium every second, releasing stupendous amounts of energy which keeps the Sun hot and bright.

SUNSPOTS AND FLARES

CLOSER INSPECTION of the solar surface often reveals groups of dark markings, or sunspots, which can be tens of thousands of kilometres across. In reality these spots are not really dark at all, but because they are around 1,500°C (2,700°F) cooler than their surroundings, they look dark against the surface. Sunspots are associated with powerful magnetic fields, which form loops above the surface. It is also from these regions that spectacular solar flares occur. In these violent eruptions magnetic energy, which has built up in the solar atmosphere, is suddenly released, with a force equivalent to millions of hydrogen bombs.

THE FUTURE

THE SUN IS half way through its life. Eventually, in around five billion years, it will begin to run out of its supply of hydrogen fuel and will expand to become a red giant. The inner planets will be destroyed and the Sun will finally shrink to a glowing ember the size of Earth, called a white dwarf.

The Moon

T HE MOON is Earth's only natural satellite, orbiting our
planet 368,000 km (228,700 miles) away. The Moon
produces no light of its own – it shines as a result of reflected
sunlight. Depending on the positions of the Sun, Moon and
Earth, the Moon goes through different phases. When the
Moon is on the opposite side of the Earth to the Sun, a full
Moon lights the night sky. A week later, the Moon has travelled
one quarter of the way around its orbit and we see just half of its
face. Another week later, a slim crescent can just be seen close to
the Sun – a new Moon.

THE LUNAR SURFACE

THE SURFACE of the Moon can be separated into two
distinct types: bright highlands and dark lunar maria, or
'seas' – as it was once thought that these dark patches were
actual bodies of water.

It is only since the start of the space age that the true
nature of the Moon has been revealed. Unlike the active Earth,
the Moon is a geological fossil. Impact craters in the bright
highlands stand shoulder to shoulder. The rocks returned by
the Apollo missions show that these highlands have changed
little since the Moon formed around 4.5 billion years ago –
apart from being pulverised by asteroid strikes during the
formation of the Solar System.

The dark maria are younger, with far fewer craters. These
vast plains were formed by huge outpourings of basalt lava,
similar in many ways to the molten rock erupted by volcanoes
on the Hawaiian Islands. These lunar eruptions occurred

between 3.8 and 2.5 billion years ago, wiping out the scars of the earlier intense asteroid bombardment.

East limb of the Moon taken by the Apollo 11 astronauts on their way back to Earth in 1969.

Mercury and Venus

MERCURY

THE INNERMOST PLANET, Mercury, never strays far in the sky from the Sun. Shortly after sunset or before sunrise, this elusive planet can occasionally be seen close to the horizon if there is no haze in the atmosphere. Very little was known about Mercury until 1974, when the Mariner 10 probe encountered this small world. Only 4,878 km (3,031 miles) across, Mercury is not much larger than our Moon, and the first images, returned by Mariner 10, revealed it to be remarkably similar in appearance. Its surface temperature ranges from - 170°C (- 274°F) at night to 350°C (662°F) during the day – the greatest contrast of any of the planets.

Mercury, the planet closest to the Sun, has a pock-marked rocky surface, scarred by billions of years of asteroid impacts .

The surface is grey in colour and is covered with impact craters. Clearly there has not been any significant geological activity on this battered world since it formed around 4.6 billion years ago.

VENUS

APART FROM the Sun and the Moon, the planet Venus is the brightest object in the sky and, at 41 million km (25 million miles) away at its closest approach, is the nearest planet to the Earth. Through a telescope, however, there is not much to see; the entire planet is shrouded with a thick white unbroken cloud cover, forever hiding the surface from our view. Appearances are deceptive – the clouds are not familiar water vapour, but corrosive sulphuric acid, and they are just the very top of a hellish carbon dioxide atmosphere. The first space probes of the 1960s and 1970s found the temperature at the surface to be 480°C (896°F), hot enough to melt lead, and the atmospheric pressure was 90 times that of Earth. The surface itself is dominated by signs of volcanic activity; and the first-ever image of the surface that was returned, by the Soviet Venera 9 lander in October 1975, shows scenes of shattered basaltic rock, illuminated by a dull orange sky.

Cloud-covered Venus is difficult to see in any great detail: the space probe Magellan used radar to map its surface.

Mars and Jupiter

MARS

EVEN TO THE NAKED EYE, the planet Mars is obviously red in colour. Half the diameter of Earth, Mars has a thin carbon dioxide atmosphere and a frigid climate with an average temperature of -58°C (-72.4°F).

The colour of Mars comes from iron-oxide dust, similar to rust, which covers most of the planet. Much of the surface

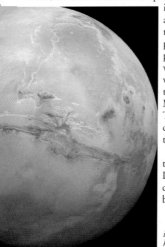

is heavily cratered, but there are many geological features that remind us of our own planet. There are bright polar caps and huge extinct volcanoes, the largest of which, Olympus Mons, is three times the height of Mount Everest on Earth. There is also an enormous canyon, which would span the USA.

Orbiting Mars are two tiny moons, Phobos and Deimos, which are almost certainly asteroids captured by Mars's gravity.

Mars is the only other planet in our Solar System so far to have raised the possibility of supporting life.

JUPITER

Jupiter is by far the biggest planet in the Solar System. In volume, it is larger than all the other planets and moons combined. Its most prominent feature is the Great Red Spot – a hurricane three times the size of Earth. This storm has raged for at least 300 years, and has probably existed much longer.

Jupiter is a very different planet to Earth; beneath the colourful, banded clouds there is no solid surface on which to stand. Instead, the mainly hydrogen atmosphere becomes thicker and thicker until it is squeezed to form a liquid ocean which extends thousands of kilometres, right to the planet's core.

Jupiter is accompanied by over 50 moons, most of which are tiny, but four of which are like planets in their own right. This miniature Solar System has been studied in great detail by the Voyager and Galileo missions, launched in 1977 and 1989 respectively.

A Voyager image of the planet Jupiter with the Great Red Spot, a storm which has been raging on its surface for hundreds of years.

Saturn and Uranus

SATURN

THE GLORIOUS RINGED planet is one of the showpieces of the Solar System. Although smaller than Jupiter, Saturn is a gas giant, with no solid surface beneath the beige clouds. Saturn is also the least dense of all the planets, with a density less than water. The ring system has a diameter of 270,000 km (168,000 miles), but a thickness of only a few tens of metres. The rings are not solid, but are composed of millions of independently orbiting chunks of ice and rock, ranging from boulders the size of cars to individual atoms.

The thousands of individual rings are thought to have formed when one of Saturn's many moons was smashed to pieces by an impacting comet, perhaps a few million years ago.

Saturn's brightness and the symmetrical perfection of its rings make it a truly unforgettable telescopic experience.

URANUS

IT WAS WHILE William Herschel was conducting a systematic survey of the sky that he discovered Uranus in 1781. Barely visible to the naked eye, this distant world shows a featureless disk in all but the most powerful telescopes. Nearly 20 times Earth's distance from the Sun, Uranus orbits in a zone of twilight, with the Sun merely a bright star. Like Jupiter and Saturn, this aquamarine world is a gas giant, although considerably smaller, at only four times the width of Earth. Its soothing colour comes from methane gas, which makes up some 15 per cent of the mainly hydrogen atmosphere. The planet, its thin coal-black rings and its family of at least 20 moons are tipped over at right angles to the Sun. What caused this is unknown, but it seems likely that, in the distant past, Uranus was struck by a planet-sized object that knocked it on to its side.

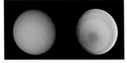

A Voyager 2 image of Uranus before (left) and after (right) strong colour enhancement.

William Herschel

German-born musician William Herschel (1738–1822) moved to Bath, UK, in 1766 and took up astronomy in 1773. He made his own telescope with which he discovered Uranus in 1781. He was George III's personal astronomer. Herschel also discovered two of Saturn's satellites.

Neptune and Pluto

NEPTUNE

TWO GERMAN astronomers, Johann Galle (1812–1910)
and Heinrich D'Arrest (1822–75) announced that they had
found a new planet on 23 September 1846. They had used the
calculations of the French astronomer Urbain Leverrier
(1819–77), who had predicted Neptune's position based on
irregularities in the orbit of Uranus. Ten days later it was
discovered that the English mathematician John Couch-Adams
(1819–92) had also independently predicted the location of the
new planet, and today Leverrier and Couch-Adams are credited
for their part in Neptune's discovery.

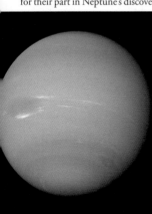

Neptune is an almost
exact twin of the planet
Uranus. Four times the
width of Earth, this gas
giant is shrouded in a blue
atmosphere of mainly
hydrogen, helium and
methane. Beneath the
atmosphere is a seething
ocean of mainly liquid
water, which extends
thousands of kilometres to
the rocky core.

*An image of Neptune showing
the Great Dark Spot, a feature
which has disappeared since
Voyager 2 took this picture
in 1989.*

PLUTO

PLUTO WAS DISCOVERED in 1930 by US astronomer
Clyde Tombaugh (1906–97) at the Lowell Observatory in
Arizona. Tombaugh used a blink comparator, an instrument
which allowed him to compare images of the same portion of
the sky taken several nights apart. After months of meticulous
searching, he found a tiny dot which moved against the stars –
Tombaugh had discovered the last world in the Sun's family of
planets. Unlike the gas giants of the outer Solar System, Pluto is
solid and, at only 2,300 km (1,430 miles) across, is smaller than
our own Moon. The Hubble Space Telescope has revealed
strange bright and dark areas with the same detail as the naked
eye can see such features on the Moon. Pluto and its large moon
Charon have yet to be visited by spacecraft. They are probably
similar to Neptune's moon Triton, with surfaces composed of
nitrogen, water and methane ices, frozen as hard as rock 5.9
billion km (3.6 billion miles) from the Sun.

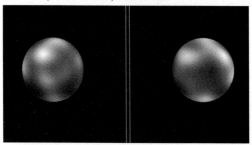

*Pluto (pictured) and its moon, Charon, closely resemble each other in size
and appearance.*

Asteroids, Comets and Meteors

IN ADDITION TO the major planets, there is a large number of other, smaller bodies which orbit the Sun. The best known are comets, 'dirty snowballs' only a few kilometres across, which originate in the depths of the Solar System beyond the orbit of Neptune. These tiny worlds are made of ice, mixed with some rocky material. Occasionally the gravitational tug of the planets forces one of these comets into an elliptical orbit that brings it close to the Sun. As it nears the Sun, the ice

begins to evaporate, and a cloud of dust and ionised gas erupts from the comet. Solar radiation pushes this cloud away from the Sun, forming a cometary tail, millions of kilometres long. It was Edmund Halley who first noticed that a comet had appeared regularly every 76 or so years, and realised it was the same object. He correctly predicted its return in 1758. The body is now known as Halley's Comet.

Halley's Comet with its spectacular tail, pictured here in 1986, appears every 76 years.

ASTEROIDS

LESS SPECTACULAR than comets are asteroids. Most of them orbit in a belt between the orbits of Mars and Jupiter. The largest is Ceres, 933 km (580 miles) in diameter, but most are tiny, with only around 20 measuring more than 200 km (125 miles) across. Asteroids are debris left over from the formation of the planets. Some are made mostly of iron – the remnants of the dense cores of worlds, which were smashed to bits by planetary collisions during the violent birth of the Solar System. Countless particles of dust also abound in the Solar System, mostly originating in the tails of comets. This dust is constantly hitting our atmosphere, and vapourising in fiery flashes known as meteors or 'shooting stars'.

Asteroid Eros is one of countless rocky fragments orbiting around the Sun, mostly between Mars and Jupiter.

Edmund Halley

Englishman Edmund Halley (1656–1742) was the Astronomer Royal from 1720 until his death. A friend of Isaac Newton, he financed the publication of Newton's *Principia Mathematica* out of his own pocket. He was also a pioneer of geophysics and contributed significantly to the evolution of the diving bell.

Stars

STARS ARE BORN from nebulae, clouds of hydrogen gas and dust in space. In the right conditions, parts of these nebulae begin to contract under their own gravity. As the cloud shrinks, the temperature soars. The gas and dust settle into a rotating disc, with a small dense core of mainly hydrogen gas, known as a 'protostar'. Astronomers believe that, in many cases, planets begin forming out of the disc. The material begins to clump together, first by collisions between the dust grains, and then due to the gravity of the growing planets.

The 'protostar' then begins to emit a strong wind of radiation that blows away most of the original shroud of gas and dust, leaving behind the young, hot planets.

Once the protostar's core reaches 10,000,000°C (18,000,000°F), the hydrogen gas begins to undergo nuclear reactions, fusing into helium. The cloud has become a star. At its core, millions of tonnes of hydrogen fuse into helium every second, keeping the star hot and bright, and stopping it from shrinking any further.

A Hubble Space Telescope image of the dusty, gaseous disc surrounding a protostar.

DIFFERENT TYPES OF STAR

NOT ALL STARS are the same. The mass of the star depends on the size of the cloud from which it formed. The more massive the star, the hotter it burns, and the temperature of its surface is shown by the star's colour. Our Sun is an average yellow star, with a surface temperature of 5,500°C (9,900°F). Stars less massive than our own are called red dwarfs, and shine

at only 3,000°C (5,450°F). The largest, hottest stars contain more mass than 100 Suns. These rare supergiants shine blue at up to 80,000°C (144,000°F) with the power of 10 million Suns.

The colour of a star reveals its temperature range and its mass.

Arthur Eddington

Arthur Eddington (1882–1944) became Professor of Astronomy at Cambridge, UK, in 1913, where he worked on the internal structure of stars. His expedition to observe a solar eclipse in 1919 provided the first proof of Einstein's theory that light from distant stars would be bent by the gravity of the Sun.

Galaxies

THE MILKY WAY

OUR SUN IS JUST one of more than 200 billion stars in our Galaxy, the Milky Way. This vast system of stars is shaped like a flattened disc, 100,000 light years across and 2,000 light years thick, with a bulge at the centre. On a clear night the faint band of the Milky Way is clearly visible. What we see is the combined light from billions of stars as we look along the plane of the Milky Way, from our position roughly half way from the centre. The Milky Way is a spiral, with stars, dust and gas clouds (nebulae) visibly concentrated along enormous spiral arms, which curve away from the galactic centre to its edge. The whole system slowly rotates, with the Sun taking 2,205 million years to complete one revolution.

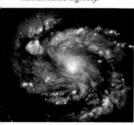

A spiral galaxy, similar to the Milky Way with spiralling arms, appears as a hazy band of bright stars across the night sky.

OTHER GALAXIES

ASTRONOMERS ESTIMATE there are around 120 billion other galaxies in the Universe, which tend to gather together in clusters. Our own cluster, called the Local Group, consists of more than two dozen galaxies, bound together by their mutual gravitational pulls. The two Magellanic Clouds, less than 200,000 light years away, are actually satellites of our own

Galaxy and are irregular in shape. From the southern hemisphere of the Earth, they are visible as bright patches in the sky.

Galaxies come in various shapes and sizes. Many galaxies are spirals, some with arms tightly wound around the centre (or nucleus) and others with a less defined, loose structure. In some spirals the arms begin from a bar-shaped feature through the nucleus. Some galaxies are irregular, with no defined shape, and others are elliptical, ranging from spherical to highly flattened.

Elliptical galaxies, like the giant M87 in Virgo (pictured), are old structures reputed to have been formed within a billion or so years of the Big Bang.

The Death of Stars

THE SUN IS ABOUT half way through its life. In around five billion years it will begin to run out of its hydrogen fuel and will expand to form a red giant, a hundred times its present diameter. The helium produced in the core will fuse into heavier elements, eventually making carbon. This phase will last just a few million years, and then the Sun's bloated outer layers will drift off into space. These layers will form a beautiful cloud called a planetary nebula, so called because they are easily mistaken for the discs of planets as seen through a telescope. At the centre, the Sun, now a white dwarf, a glowing ember the size of the Earth, will slowly cool and fade away.

VIOLENT ENDS

SOME STARS, much more massive and luminous than the Sun, quickly exhaust their nuclear fuel in just a few million years. In their cores, the temperature can reach billions of degrees, allowing the carbon produced by nuclear fusion to keep fusing into heavier elements. Eventually iron is formed, which halts the nuclear reactions. Then the star immediately collapses and blows itself to smithereens in a colossal explosion called a supernova. The tiny remnant left behind is a neutron star, a sphere the size of a large city, a spoonful of which would weigh half a billion tons.

A more bizarre fate awaits the most massive stars. Once their fuel has run out and they begin to collapse, their gravity is so great that they continue to shrink past the neutron star stage. As they become smaller and denser, their gravity increases dramatically until not even light can escape. They have become black holes.

When a star explodes, it blows off most of its gaseous mass, creating a bright supernova (pictured).

EXPLORING SPACE

Early Rockets

PRIMITIVE ROCKETS

ROCKETS WORK using a very basic principle. As a rocket's fuel burns, it produces expanding gases, which escape through one end of the rocket. In the same way that a fired rifle 'kicks' backwards, the escaping gases propel the rocket forward.

The principle of rocket power was used in warfare before it was applied to space travel. During the British campaign in India in the late eighteenth century, the defending troops used rockets with a range of up to 2.5 km (1.6 miles) to wreak havoc on the advancing British army. In the early nineteenth century, the British used larger rockets to firebomb cities such as Copenhagen and Boulogne.

Konstantin Tsiolkovsky

Konstantin Tsiolkovsky (1857–1935) is considered to be one of the founders of astronautics. The son of a Polish forester, he theorised about space travel decades before anyone else. His most famous quotation is from a letter in 1911: 'The Earth is the cradle of the mind, but we cannot live forever in a cradle.'

ABOVE THE ATMOSPHERE

IN 1903, the Russian rocket pioneer Konstantin Tsiolkovsky (1857–1935) proposed replacing the solid, gunpowder-derived rocket fuels with a more efficient liquid propellant, realising that rockets could, theoretically, be used to travel above the atmosphere and into space. All substances require oxygen in which to burn. With no oxygen in space, Tsiolkovsky suggested that a rocket carry its own supply, to burn with the liquid-hydrogen fuel.

THE V-2

IT WAS NOT UNTIL the Second World War that rocket science really took off: led by Werner von Braun (1912–77), a German team of engineers and scientists built the V-2 missile. This rocket carried 900 kg (2,000 lbs) of explosives and could travel from a variety of launch sites on mainland Europe and rain destruction on the city of London. At the end of the war, von Braun and his team transferred to the USA, along with a hoard of V-2 missiles. They modified and improved the V-2 in a race with Soviet engineers who were developing their own rockets. This competition led to travel in space and to robotic missions to the planets.

Robert Goddard

American Robert Goddard (1882–1945) built and launched the first liquid fuelled rocket, which carried a small camera, barometer and thermometer to perform basic scientific experiments. His pioneering work on rocket propulsion coincided with Germany's development of V-2 ballistic missiles.

Rocket Launchers

SINCE THE HEADY DAYS of the first launches into space in the late 1950s, the size and power of rockets have increased dramatically. The Soviet Union, the US, and later Europe, have invested heavily in the development of boosters which can haul heavier payloads into Earth orbit.

SATURN V

THE MOST FAMOUS rocket is the Saturn V, which sent astronauts to the Moon during the Apollo programme in the 1960s and 70s. This 110 m- (361 ft-) high rocket generated as much energy on lift-off as is required to light up New York City for over an hour. The intense roar of the launch in Florida was picked up by seismometers in New York. The Saturn V performed without significant problems for any of its 13 launches.

PROTON

ORIGINALLY DEVELOPED in the 1960s as a nuclear missile, the Proton rocket is the workhorse of the Soviet space launchers. This three-stage, heavy duty booster can place over 20 tonnes into low Earth orbit, and would probably have been used to its full capacity if the Soviets had proceeded with their own manned Moon programme.

Ariane 5, capable of placing both satellites and manned missions into Earth orbit, was the European Space Agency's answer to NASA rockets.

ARIANE 5

THE EUROPEAN Space Agency's (ESA's) Ariane rocket was originally planned to send both satellites and manned missions into Earth orbit, but its own manned programme has yet to begin. The latest model, Ariane 5, has undergone many improvements and modifications, and can place two 3-tonne satellites or a single 6.8-tonne satellite in orbit.

DELTA

THE DELTA II rocket is the most commonly used American rocket, launching various commercial and scientific unmanned missions, more than a dozen times each year. The Delta II can place satellites weighing around five tonnes into low Earth orbit.

The most famous US rocket: Saturn V, used during the Apollo programme, launches from Cape Canaveral.

Werner von Braun

A rocket engineer in Hitler's Germany, von Braun (1912–77) developed the V-2 rockets. At the end of the Second World War he surrendered to the Americans, who wanted to acquire his advanced rocketry skills. As the space race got underway, his technology was put to use in making rockets to launch satellites into space, including the Saturn rocket used in the Apollo 11 Moon landing in 1969.

Launch Sites

THE BUSINESS OF launching a rocket from the ground up into space is no simple matter. The thousands of personnel and highly specialised equipment necessary mean that launch sites are vast sprawling technological centres.

CAPE CANAVERAL

THE MOST famous launch site is the John F. Kennedy Space Center at Cape Canaveral, Florida, USA. The first launch at the centre was of a captured German V-2 rocket on 24 July 1950, and it has been the focus of the US space programme ever since, as well as being the primary landing site for the space shuttle fleet. Once a spacecraft has been launched, control is handed over to mission control at the Johnson Space Center, Houston, or to the Jet Propulsion Laboratory, California, for unmanned planetary missions.

One of the Kennedy Space Center's most prominent features is the Vehicle Assembly Building (VAB). Standing 160 m (525 ft) high and covering an area of three hectares (eight acres), the VAB is one of the largest buildings in the world. This mammoth construction was originally used to assemble the Saturn V rockets used in the Apollo programme, and now services the space shuttle.

BAIKONUR

THE RUSSIAN SPACE programme's main launch facility is the Baikonur Cosmodrome. It came into existence in 1955 under a veil of secrecy. The whole cosmodrome covers an area 85 by 125 km (53 by 78 miles) in the hostile environment of western Kazakhstan. From here Yuri Gagarin (1934–68) began the historic first manned space mission on 12 April 1961, and many of spaceflight's other historic moments have begun here, too. Since the collapse of the Soviet Union, Baikonur has suffered major economic problems and its long-term future is uncertain.

KOUROU

THE EUROPEAN SPACE agency, ESA, launches its rockets from Kourou in French Guyana. As there are no unpopulated areas in Europe itself, the French Space Agency, CNES, decided to construct its own launch site on the South American subcontinent in 1964. Since then, the complex has become the gateway to space for the whole of the ESA, which also launches some of its spacecraft from Russia's Baikonur Cosmodrome.

The best-known launch site in the world: the John F. Kennedy Space Center at Cape Canaveral, Florida, USA.

Theory of Orbits

IMAGINE AN IMPOSSIBLY high mountain, with its summit above the Earth's atmosphere. A shell fired horizontally from a powerful cannon on the mountain top will eventually fall to the ground. But if the cannon can shoot its shell fast enough, the curved surface of the Earth will drop away beneath it at the same rate as the shell is falling towards it – it is in orbit. This only works above the atmosphere, whose friction would slow the projectile, bringing it down. To orbit the Earth, a spacecraft must be travelling at 27,350 kph (16,990 mph) and at this speed, it will circle the world every 90 minutes.

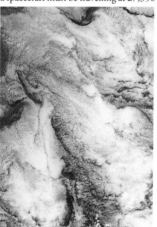

When a rocket is launched, carrying a satellite or spacecraft into orbit, it begins its journey pointing straight upwards, so it can quickly leave the thick, lower atmosphere. Then its path begins to trace an arc. Only a few minutes later it is following an orbital path, parallel to the Earth's curved

A view of Earth from a satellite orbiting the planet.

surface, at an altitude of around 200 km (124 miles). Higher than this, the pull of Earth's gravity is reduced – the rocket has to propel the spacecraft higher, but its final speed is lower. For example, the Moon is 376,000 km (233,650 miles) from Earth and orbits our planet in 27 days.

Most of the mass of a rocket on a launch pad is fuel. As this is rapidly consumed during the trip to orbit, the lower stages of the rocket are shed until just the spacecraft remains.

Earth and satellites

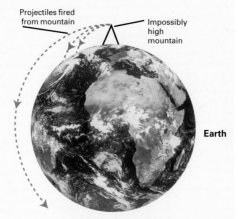

Projectiles fired from mountain

Impossibly high mountain

Earth

In orbit – if fired fast enough

Diagram explaining how satellites can orbit around Earth.

Satellites

WATCHING THE WEATHER

SATELLITES ORBITING THE EARTH have a variety
of uses. As well as mapping and monitoring our planet, they are
used extensively to monitor the world's weather from above.
The first weather satellite, Tiros 1, was launched in April 1960.

Its first fuzzy black and white image led to a revolution in weather prediction. Meteorologists can now have advance warning of severe weather, as real-time images allow them to see developing storms and hurricanes around the world. Some weather satellites orbit close to Earth, providing regional detailed information on the weather. Others orbit at a distance of 35,780 km (22,234 miles) above the Earth's equator. This far from Earth, one orbit takes exactly 24 hours, the same time it takes for the Earth to rotate once. This allows the satellite to remain fixed over the same place, allowing steady, global views of the weather.

NAVIGATION AND COMMUNICATION.

SATELLITES ALSO provide pinpoint accuracy for modern navigation. The Global Positioning System (GPS) is a network of more than 24 satellites in different orbits around the Earth. Portable receivers on the ground track their distance to several of these satellites at once, allowing them to know their precise location on the Earth, to within a few metres. As well as providing essential navigation for ships and aircraft, many explorers and hikers now carry hand-held GPS receivers, and some new cars are equipped with them.

Before the first satellites were launched, long-distance communications relied on cables, stretching across continents and oceans. Today we can receive telephone calls and television and radio signals, which have been sent up into space from a remote location on Earth, and bounced back to us via orbiting satellites.

A satellite picture of a hurricane, clearly showing its epicentre.

Probing Space

One of the earliest images of space-probing: a lunar panorama photographed by Luna 9 in 1966.

FIRST STOP: THE MOON

THE VERY FIRST planetary space probe was the Soviet Luna 1, in January 1959. Although it missed the Moon by almost 6,000 km (3,700 miles), it proved the Moon had no magnetic field before falling silent 62 hours after launch. The Soviets then managed to make contact with the Moon in September 1959, with Luna 2, which crash-landed. Luna 3, the following month, passed around the far side of the Moon and photographed parts of the surface never seen by human eyes before. Luna 9 touched down safely in 1966, returning the first-ever pictures from the surface.

The Americans were not far behind, with their Ranger probes operating between 1961 to 1965. The first six all failed, but the last three returned detailed pictures of the lunar surface right up to the last second before impacting on the Moon.

ON TO THE PLANETS

MEANWHILE, the first missions to the other planets were underway. Again, both the USSR and USA had failures on their first attempts. Contact was lost with Venus-bound Venera 1, 7.5 million km (5 million miles) from Earth, and Mariner 1 also failed to reach Venus, falling into the ocean shortly after take off.

Mars proved an elusive target, especially for the Soviets. From their first attempt in 1961 right through to the loss of the Mars 96 probe in 1996, none of their 11 Mars probes have been fully successful. Soviet and US scientists learned from their mistakes fast. Both rockets and spacecraft became more robust, and probes soon began to revolutionise our knowledge of the planets.

Planetary Probe Programmes

PIONEER

NASA'S PIONEER programme consisted of 13 missions, launched from 1958 to 1978. The first three failed, but Pioneers 4 to 9 provided valuable data on the Sun. The last two missions studied Venus, but the most famous were Pioneers 10 and 11, which blazed a trail to Jupiter and Saturn, in preparation for the more complex Voyager missions.

VENERA

THE SOVIET VENERA programme was a series of 16 probes to Venus, launched between 1961 and 1983. Although the first missions failed, Veneras 7 to 14 landed safely, analysing the soil and taking the first pictures of Venus's surface. These sturdy

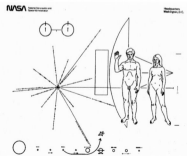

This image appears on the side of Pioneers 10 and 11, providing would-be visitors with directions to Earth and indicating what they would find there.

probes did not last long in the hostile Venusian environment.
Venera 11 managed to transmit for 95 minutes after touchdown.

MARINER

THE PLANET MARS was the destination for Mariners 4, 6
and 7 flyby missions, in 1965 and 1969. Between them they
returned 223 images, mainly of the southern Martian
hemisphere, which revealed only vast plains covered with impact
craters. Mariner 9 became the first spacecraft to orbit Mars in
1971. During its year circling Mars, it discovered the red planet
to be much more interesting than its predecessors had suggested.
In its 7,329 images were the first-ever views of the giant Martian
volcanoes and the huge canyon Valles Marineris.

VIKING

THE VIKING MISSIONS are
among the most ambitious space
probes to date. Two spacecraft
entered Martian orbit in 1976
and began an extensive mapping
programme. Meanwhile, a lander
separated from each orbiter and
successfully touched down,
providing our first-ever view of
the Martian landscape. Robot
arms scooped up soil samples
and analysed them for life, but
the soil proved to be barren.

*A view of the Martian landscape taken
during the Viking missions in 1976.*

The Voyager Missions

IN 1977, NASA took advantage of a rare alignment of the outer planets to launch its most ambitious unmanned planetary expedition. The two Voyager spacecraft would fly past Jupiter and Saturn, and Voyager 2 would continue on to both Uranus and Neptune.

One of the Voyager spacecraft.

JUPITER

DURING THE encounters with Jupiter in 1979, mission scientists were amazed at the turbulence visible in its atmosphere and Great Red Spot. Voyager also discovered great variety on the surfaces of the four largest moons: Callisto and Ganymede were ancient and heavily cratered, but Europa and Io were altogether different. Icy Europa was smooth and covered in huge cracks, like an eggshell, resembling an ocean with a frozen crust. Io was alive with huge active volcanic eruptions – the first ever seen off the Earth – spewing gas and ash hundreds of kilometres above its surface.

SATURN

VOYAGER 1 REACHED Saturn in November 1980; Voyager 2 in August 1981. Their cameras revealed ultra-fine detail in the ring system and discovered complex molecules in the atmosphere of the mysterious moon, Titan.

URANUS

THE GREEN CLOUDS of Uranus showed very little detail to Voyager 2's cameras, during its encounter in 1986. The whole planet was shrouded in a high-altitude haze. However, Voyager did discover 10 satellites ranging from 26 km to 154 km (16 to 96 miles) in diameter that had not been seen before.

NEPTUNE

IN AUGUST 1989 Voyager 2 closed in on Neptune. The deep blue atmosphere showed far more detail than the bland disc of

Uranus. There was a great dark blue hurricane, the size of Earth, similar to the Great Red Spot on Jupiter. Voyager 2 also discovered nitrogen geysers, 8 km (5 miles) high, erupting from the frigid surface of one of Neptune's eight known moons, Triton.

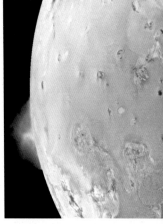

Over 20 years after their launch, both Voyagers are still operational and sending back data from the very fringes of the Solar System, as they head out towards interstellar space.

A volcanic eruption from Io, one of Jupiter's moons, as recorded by Voyager.

Giotto – Exploring Halley's Comet

THE MOST FAMOUS of all comets came close to the Earth in 1910 before heading back into the outer Solar System on its 76-year elliptical path around the Sun. On its return in 1986, the ESA planned a daring mission to send a probe hurtling through its head of gas and dust to its nucleus, to study a comet from close-up for the very first time.

A SUICIDE PLUNGE

THE GIOTTO spacecraft was a squat, barrel-shaped probe 1.85 m (6 ft) across and 2.85 m (9 ft) long. On board was an array of science experiments including a camera, which would image the mysterious icy nucleus and witness the process of its gradual evaporation as it approached the Sun.

On 13 March 1986, with its dust-shield armour facing forwards, Giotto plunged towards Halley's Comet at a velocity of nearly 80 kps (50 mps) relative to the nucleus. On-board detectors

A close up view of the nucleus of Halley's Comet, from Giotto. Bright jets of gas and craters are visible.

began registering impacts from tiny particles streaming away from the nucleus, and the camera sent back ethereal images of the potato-shaped 15 km- (9 mile-) long nucleus. Hills, craters and patches of differing brightness were clearly visible, as well as jets of gas and dust – the source of the tail, which stretched more than 6 million km (3.7 million miles) behind the nucleus.

Giotto passed just 596 km (370 miles) from the nucleus, but 14 seconds before closest approach, a rice-sized particle struck the probe, causing it to wobble. Other instruments continued to function but the impact had caused the camera to fail. Its last image, from a distance of 1,675 km (1,040 miles), revealed surface features 100 m (328 ft) across on the nucleus.

Giotto contained state-of-the-art equipment with which to discover more about the icy heart of Halley's Comet.

Magellan and Galileo

MAGELLAN

THE MAGELLAN CRAFT was carried into space on board the space shuttle in May 1989. After a 15-month interplanetary cruise, it arrived at Venus and went into orbit. Its mission was to map the whole surface of Venus in fine detail, but Magellan carried no camera on board. The thick, choking atmosphere with its white clouds allowed no view of the surface, so Magellan mapped the planet using radar. From its orbit, 289 km (180 miles) above the planet, Magellan sent radar pulses down to the surface. The reflected signals gradually built up a picture of the surface strip by strip across the planet. The radar images from Magellan revealed a multitude of volcanic features. Shield volcanoes and lava channels were accompanied by huge coronae, vast, fractured circular features possibly formed above huge bodies of molten magma. Magellan continued imaging features as small as 100 m (330 ft) across for four years, before burning up in the Venusian atmosphere.

Volcanic channels on Venus, as captured by Magellan.

GALILEO

LAUNCHED IN October 1989, the Galileo orbiter arrived at Jupiter six years later and began an exhaustive survey of the giant planet and its moons. While approaching Jupiter, Galileo released a probe that plunged into its atmosphere. After being slowed by a heat shield, the probe descended by parachute, analysing the alien atmosphere for almost one hour before being crushed by the pressure 100 km (62 miles) below the cloud tops. Jupiter's four planet-sized moons were particular mission targets. Galileo monitored the immense volcanic eruptions on Io, first seen by the Voyager encounters over two decades earlier. Galileo also found evidence for a liquid ocean of water beneath the cracked icy shell of Europa.

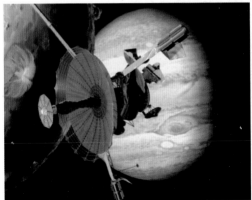

Galileo during a fly-by of Io with Jupiter in the background.

Return to Mars

IN DECEMBER 1996 NASA launched Pathfinder, the first lander to visit Mars since the Viking missions 20 years earlier. Pathfinder represented a new initiative of smaller, more focused probes. On 4 July 1997 Pathfinder entered the Martian atmosphere and dropped to the surface, cushioned by air bags that deflated before the probe began its survey of the surface.

Its stereo camera extended to a height of 1 m (3.2 ft) and it snapped pictures of the surrounding Martian landscape. Pathfinder also had on board the Sojourner rover. This was no larger than a microwave oven, and explored the vicinity of the lander and analysed the rocks deposited by the ancient flood. Designed to last for only one month, the mission continued to function for 83 days.

In 1997 Mars Global Surveyor (MGS) arrived at Mars and began an extensive survey from orbit. As well as a camera that can image individual boulders, MGS is equipped with sensors to determine the chemical composition and the exact elevations of features on the surface. Its most exciting discovery so far has been a series of small channels or stream-like features that appear

The Sojourner rover explores Mars, analysing the boulder nicknamed 'Yogi'.

to have been formed in the last million years or so (which is very recently in astronomical terms) by liquid water erupting from underground reservoirs.

For a planet thought to be completely barren, this discovery altered our perception of Mars. In 2001 the Mars Odyssey was launched with the intention of determining whether the environment on Mars was ever conducive to life and to study potential radiation hazards to possible astronaut missions.

Most recently, in 2003–4, the Mars Exploration Rovers (MERs) 'Spirit' and 'Opportunity' landed on Mars in order to help determine if life ever arose there, to characterize the geology of the planet and to prepare for future human exploration. The extraordinary pictures they have sent back have told us much about this mysterious red planet.

Fresh gullies on Mars's surface discovered in July 2000 by MGS.

Ulysses and SOHO

EXPLORING THE SUN

SINCE 1960, more than 40 spacecraft have studied the Sun, from orbit around both the Sun and the Earth. All these missions have concentrated their science on the equatorial regions of the Sun. The Ulysses probe is the first spacecraft to study its poles. In order to achieve polar orbit, the probe had to travel to Jupiter first, whose immense gravity would deflect its path downwards out of the plane of the Solar System and back towards the Sun. In this six-year-long polar orbit, Ulysses passed directly beneath the Sun in 1994 and obtained the first clear view of the Sun's south pole. The results were a surprise: astronomers had expected the strength of the magnetic pole to be intense, but Ulysses showed that it was ill-defined and vague. The following year, the spacecraft passed over the Sun's north pole and took further measurements of the magnetic fields and radiation.

SOHO

THE EUROPEAN Space Agency's Solar and Heliospheric Observatory (SOHO) was launched in 1995. It observes the Sun constantly from its vantage point 1.5 million km (1 million miles) closer to the Sun than the Earth. Here the gravitational pulls of the Sun and Earth cancel each other out, so the spacecraft remains fixed in position.

SOHO is providing our best uninterrupted view of the Sun, at many different wavelengths of light. Its instruments have recorded gigantic tornadoes in the Sun's atmosphere and bizarre rivers of searing plasma (a gas in which the atoms have had their

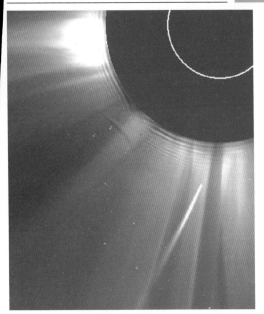

Image taken by SOHO of a comet plunging into the Sun.

electrons stripped away and have become ionised) which wrap around the Sun. SOHO is also helping astronomers predict bad 'space weather', caused by huge eruptions of particles from the Sun interfering with electronics on Earth.

MAN IN SPACE

A Fascination with Space

MYTHS AND LEGENDS OF HUMANS travelling to the Moon or the stars are probably as old as language. In AD 160, Lucian of Samosata (*c.* AD 117–180) in Greece wrote one of the earliest surviving accounts of such a voyage, in which a ship is caught by a freak storm at sea, and sent on a week-long adventure to the Moon. The astronomer Johannes Kepler imagined the effects of zero gravity in his book *Somnorium* ('Dream'), one of the first works of science fiction, published in 1634.

A crowd of press and public gather to watch the launch of Apollo 7 in 1967.

ONE SMALL STEP

THE ANCIENTS could only see the Universe with the naked eye, but when Galileo Galilei pointed his new telescope at the sky in 1609, the details it revealed served to further heighten humanity's fascination with the heavens. Once the first German V-2 rockets had rained down on England during the Second World War, scientists and engineers realised the potential of the rockets to launch humans into space. This was finally achieved in April 1961, when Yuri Gagarin orbited Earth in his Vostok 1 capsule.

Eight years later, in July 1969, the fantasies of people throughout history were finally realised when Neil Armstrong (b. 1930) set foot on the lunar surface as the whole world watched.

Since then, man's fascination with space continues to be just as strong, spurred on by technological advances which promise more widespread access to space. The numerous films set in space that are produced every year bear witness to this, and science fiction books are ever popular. Many other books and television programmes also exploit our fascination with other life forms in space – the myth of 'little green men' is a difficult one to quash!

The whole world watched as Neil Armstrong became the first man to set foot on the Moon.

The First Satellites

SPUTNIK 1

THE FIRST VICTORY of the space race was won by the
Soviet Union. On 4 October 1957 the Soviets announced
to the world that they had successfully launched the very first
artificial satellite into orbit around the Earth. The world was
thrilled – few had expected the USSR to beat the technologically
sophisticated USA to such a significant milestone. The satellite
was an 83-kg (183-lb) aluminium sphere, radiating four radio
antennae. Its elliptical orbit took it as far as 945 km (587 miles)
from the Earth, and just 227 km (141 miles) at its closest
approach. It functioned for
21 days and, powered by a
one-watt transmitter, its
characteristic beep-beep was
heard all over the world. Many
Americans were terrified: if the
Soviets could send Sputnik
into orbit, then why not a
nuclear warhead?

Sputnik 1.

LAIKA

NO SOONER had the world
recovered from the launch of
Sputnik 1, than the Soviets
accomplished their next
milestone – within a month.
On 3 November 1957 Sputnik 2
was sent into orbit. At 508 kg

(1,120 lbs), it was more than six times heavier than its predecessor, and it carried the first life form into orbit. Laika the dog was inside a small capsule which contained a life-support system, and the satellite stayed in orbit for almost 200 days. Sputnik 2 was not designed to be recovered, and Laika (Russian for 'barker') captured the hearts of people around the world as she perished a few days into her flight, due to high temperatures in the spacecraft.

Laika, on board Sputnik 2, the first living creature to be launched into orbit.

First Human in Space

AFTER THE TRIUMPH of Sputnik 1 in 1957, the Soviets achieved another spectacular first with the successful launch of Vostok 1 on 12 April 1961. This spacecraft carried on board a 27-year-old man named Yuri Gagarin, who began the mission with an exuberant cry of 'Poyekhali!' ('Let's go!'). This was the first time in history that a human being had left the protection of the Earth and entered the great unknown of space. The Vostok spacecraft was based on the Kosmos unmanned spy satellites, which would re-enter the atmosphere

Yuri Gagarin is greeted by army officials upon his return from the Vostok 1 space mission.

once their photographic film was used up, and be retrieved. Gagarin's spherical capsule was only 2.3 m (7.5 ft) across and had a separate equipment module attached, containing the communications package necessary for contact with mission control. As most of the controls of the spacecraft were automated or controlled from the ground, Gagarin was not required to pilot his craft. But in case of an emergency, he was given a sealed envelope that contained the necessary codes to give him manual control.

A SOVIET TRIUMPH

THE MISSION lasted just 108 minutes, during which Vostok 1 completed one orbit of the Earth and reached a maximum altitude of 327 km (203 miles) above the surface. Then the spacecraft automatically fired its retro-rockets, which slowed it down enough to re-enter the atmosphere. The heat-shield performed as planned and Gagarin ejected from the capsule at an altitude of a few kilometres, landing back in the Soviet Union by parachute much to the surprise of the local peasants! Gagarin was now a national hero, and the USA became increasingly concerned that it was lagging behind in the space race.

Yuri Gagarin

Russian cosmonaut Yuri Gagarin (1934–68) was the first man to travel in space, making one complete orbit of the Earth in 1961 in the Vostok 1 space satellite. He became head of cosmonaut training, and was killed when his jet aircraft crashed in 1968. He was the first hero of space travel, receiving honour worldwide. In the USSR, the town of Gzhatsk was renamed Gagarin.

Project Mercury

IN 1959, NASA began working on their first manned space missions, Project Mercury. Seven test pilots were chosen to train as 'astronauts' who would travel into space, launched on a Redstone or Atlas rocket. At 2 m by 1.9 m (7 ft by 6.2 ft), the bell-shaped Mercury capsule was barely large enough for a single person.

The first flights were unmanned and in January 1961, Ham the chimpanzee, was sent to an altitude of 253 km (157 miles) in a Mercury capsule, proving to NASA, upon his return, that living things could survive in space.

Only 23 days after Yuri Gagarin became the first person in space, NASA was ready to send up the Freedom 7 Mercury spacecraft, piloted by Alan Shepard (1923–98). His Redstone rocket was not powerful enough to reach orbit, so Freedom 7 travelled into space in an arc, reaching an altitude of 187 km (116 miles), and a velocity of 8,282 kph (5,145 mph). Shepard splashed down in the ocean, 487 km (302 miles) from Cape Canaveral, after a flight lasting just 15 minutes.

INTO EARTH'S ORBIT

AFTER AMERICAN Virgil Grissom (1926–67) completed another sub-orbital flight in July 1961, fellow countryman John Glenn (b. 1921) flew his Mercury capsule into orbit, launched by a more powerful Atlas booster. His flight lasted five hours, during which he tested the controls of his capsule and completed three orbits of the Earth. Three more Mercury flights followed, culminating in Gordon Cooper's (b. 1927) 34-hour flight in May 1963. By the end of his flight, virtually every system in his capsule had failed, forcing Cooper to attempt a manual entry back through the atmosphere. Cooper not only survived, but also managed to splash down within sight of the rescue ship.

John Glenn orbited Earth for five hours in his Mercury capsule.

John Glenn

Born in 1921 in Ohio, USA, John Glenn flew 59 combat missions during the Second World War. He became the first American to orbit the Earth in his Friendship 7 capsule on 20 February 1962. Glenn returned to space in October 1998 on board the space shuttle, at the age of 76.

The Space Race

A TECHNOLOGICAL FLEXING OF MUSCLES

FROM THE ASHES of the Second World War, a very different kind of conflict arose. The US and the USSR, with their conflicting politics and ideals, embarked on a struggle to show their superiority by developing the best weapons and technology. In this 'Cold War', the showcase was the space race. From the 1950s, the USA and the USSR strived to be the first in space and the cost was no barrier.

Although the Soviets led the way by launching the first artificial satellite, Sputnik, in 1957, and the first man, Yuri Gagarin, in 1961, the Americans began to catch up rapidly. Across the USA, hundreds of thousands of people were working on the space programme for the new US space agency, NASA. In 1961, President John F. Kennedy publicly announced the USA's intention to send a manned flight to the Moon by the end of the decade and it was NASA's job to make it happen.

Major Yuri Gagarin at a ceremony in his honour, in April 1961, shortly after he became the first man in space.

This technological flexing of muscles extended out into the Solar System, as both nations began sending probes to the planets, but the ultimate goal was the Moon. Soon the USA and USSR were launching two-man crews into Earth orbit, where they practised with all the equipment and techniques necessary for a Moon landing.

THE END OF AN ERA

THE SPACE RACE was lost by the Soviets when the first manned landing on the Moon was achieved by Apollo 11 in July 1969. The Soviets never followed. After five more landings, NASA cancelled the Moon programme after Apollo 17, in 1972. Today, the USA and Russia maintain space programmes, but on a smaller scale than during the height of the space race.

Buzz Aldrin salutes the Stars and Stripes after planting the flag on the Moon in July 1969.

The Soyuz Programme

THE SOYUZ SPACECRAFT consists of three modules, which give it an almost insect-like appearance. At the front is the orbital module, where cosmonauts work once the Soyuz has reached orbit. The equipment module at the rear contains the rocket engines and fuel, as well as supporting the solar panels. The orbital module is equipped with a robust docking mechanism and it can also depressurise to become an airlock, allowing cosmonauts to leave the spacecraft to carry out a spacewalk.

When a mission is complete, the spacecraft separates and the middle of the Soyuz, the descent module, returns the crew to Earth. This basic design has been the workhorse of the Soviet/Russian space programme since the first mission, Soyuz 1, in 1967.

Soyuz in orbit around the Earth.

APOLLO SOYUZ

ONE OF THE highlights of the Soyuz programme was the Apollo–Soyuz test project, the first multi-nation spaceflight. In July 1975, Soviet cosmonauts Alexei Leonov (b. 1934) and Valery Kubasov docked their Soyuz craft with an Apollo craft carrying US astronauts Donald 'Deke' Slayton (1924–93) and Thomas Stafford (b. 1930). Once the hatches were opened, Leonov and Slayton shook hands in what is one of the most famous images of the space race.

Alexei Leonov and Donald Slayton greet each other during the Apollo Soyuz test.

Soyuz has proven to be one of the most versatile and reliable spacecraft of all time. As well as a space-station ferry, it has been used on long-duration missions and in scientific Earth observation programmes. Over 80 Soyuz spacecraft have been into space so far and the latest upgraded models are still used to ferry cosmonauts to and from the Mir space station.

Valentina Tereshkova

Born in 1937, Valentina Tereshkova was selected as a Soviet cosmonaut due to her parachute experience. In 1963 she became the first woman in space. The Vostok 6 mission, her only spaceflight, lasted nearly three days, and completed 48 orbits of the Earth.

Gemini

FOLLOWING THE HIGHLY successful Mercury programme, the Gemini missions were instigated to develop and test all the routines necessary for a lunar mission. Between April 1964 and November 1966 NASA

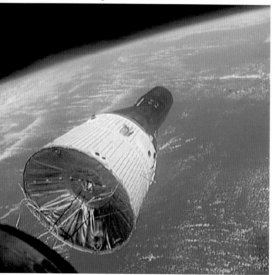

Gemini 7, as seen by the crew of Gemini 6.

launched 12 Gemini spacecraft into Earth orbit. The first two were unmanned tests, but the other 10 each carried two astronauts on spaceflights crammed with tests and procedures.

TECHNOLOGICAL HURDLES

AFTER AMERICAN Edward White's (1930–67) spacewalk during the Gemini 4 mission, NASA decided to attempt two milestones in one double mission. If Apollo was to make it to the Moon, astronauts would have to prove they could stay in space for over a week, and rendezvous two vehicles in orbit. In December 1965, Americans Frank Borman (b. 1928) and James Lovell (b. 1928) spent 14 days circling Earth on board Gemini 7 in a space no larger than a modest lavatory, evaluating the performance of both the spacecraft and the human body. Eleven days into this gruelling mission, Borman and Lovell were met in orbit by fellow Americans Walter Shirra (b. 1923) and Thomas Stafford, on board Gemini 6. In an impressive display of formation flying, Shirra brought his spacecraft to within one metre of Gemini 7. While travelling at over 27,300 kph (17,000 mph) the two crews were able to wave at one another through their windows.

SPACE DOCKING

THE NEXT STEP was to attempt an actual docking in space. Geminis 8, 10, 11 and 12 docked with unmanned Agena rocket stages, lofted into orbit before each mission. During the final Gemini flight, Edwin 'Buzz' Aldrin (b. 1930) spent a record five-and-a-half hours outside his spacecraft. He simulated making repairs to the capsule and proved it was possible for humans to work in space itself. With all the goals of Gemini achieved, NASA was ready to begin the Apollo programme.

124 MAN IN SPACE

The First Spacewalk

ALEXEI LEONOV

COSMONAUTS ALEXEI LEONOV and Pavel Belyayev
(1925–70) lifted off into space on 18 March 1965. Their
Voskhod 2 mission was the Soviet's sixth manned mission and
its goal was to send a human being outside the sanctuary
of his spacecraft in the very first spacewalk. During Voskhod's
second orbit of the Earth, Alexei Leonov, wearing his bulky
spacesuit with its own oxygen supply, entered an inflatable
airlock and sealed the hatch behind him. Having depressurised
the airlock he drifted outside into space. Leonov floated in
space for 12 minutes, attached to his ship by a 5.3 m- (17.4 ft-)
long safety line. Despite predictions of nausea or vertigo,
Leonov felt fine and was able to take photographs of his
spacecraft and the Earth. There was a tense moment when he
found he could not fit back in the airlock but, staying calm, he
allowed some of the oxygen in his suit to bleed into space, and
managed to squeeze back inside the Voskhod capsule. A
malfunction in the ship's life-support system forced Belyayev
and Leonov to re-enter the Earth's atmosphere sooner than
planned, and they landed way off target, waiting alone for three
days in the Ural mountains before being rescued.

GEMINI 4

LESS THAN three months later the Americans accomplished
their own first spacewalk. On 3 June 1965 Edward White opened
up the hatch of his Gemini 4 spacecraft and spent half an hour

*US astronaut Ed White takes the plunge, becoming the first American
to walk in space.*

outside in space. He floated around attached by an oxygen line, manoeuvring using a hand-held gun that fired nitrogen gas. Enraptured by the experience, White declared that it was the saddest moment of his life when he ended his spacewalk.

The Apollo Programme

WITH THE EXPERIENCE gained from the Mercury and Gemini missions, NASA was ready for the Apollo Moon programme. The spacecraft that was to be launched by the enormous Saturn 5 rocket consisted of two separate vehicles, the command module and the fragile lunar module which would land on the Moon. Once en route, the astronauts in the command module would separate from the final stage of the spent booster, turn around, and dock with the lunar module. The booster would be discarded and the joined spacecraft would continue to the Moon. Four days later and 376,000 km (233,650 miles) from Earth, the Apollo craft would enter lunar

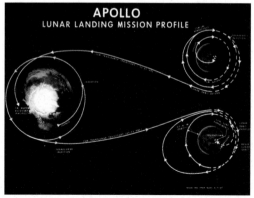

NASA diagram of the Apollo lunar landing mission.

orbit. One astronaut would stay in the command module while the other two would descend in the lunar module. Once the surface mission was complete, the upper portion of the lander would leave the Moon and return the two moonwalkers to the command module, which would then head back to Earth.

APOLLO 8

ON APOLLOS 7 and 9, astronauts tested the spacecraft in Earth orbit. But NASA, fearing a Soviet lunar attempt, decided to send Apollo 8 into lunar orbit earlier than planned. On Christmas Eve 1968, James Lovell, Frank Borman and William Anders (b. 1933) fired their single engine and slowed down enough to be captured by the Moon's gravity. They completed 10 orbits of the Moon, photographing the desolate view from only 60 nautical miles (111 km/69 miles) above the cratered landscape. But it was the sight of the distant Earth, rising above the lunar horizon, that was mesmerising. For the first time, humans had witnessed their home planet, framed by the foreground of another heavenly body.

Earthrise as seen from Apollo 8.

Apollo 11

NEIL ARMSTRONG and his crew, Michael Collins (b. 1930) and Edwin 'Buzz' Aldrin, were all veterans from the Gemini missions. Their task was the ultimate test in space-flight: to land on the Moon's surface and return safely to Earth.

Apollo 11 lifted off from Cape Canaveral on 16 July 1969 and entered lunar orbit four days later after a flawless trans-lunar flight. The tension was high at mission control in Houston, as Apollo 11 was out of radio contact, on the far side of the Moon, while Armstrong undocked the lunar module Eagle from the command module Columbia. Armstrong and Aldrin descended towards the surface, despite Eagle's computer controls overloading six times. Then, at an altitude of 610 m (2,001 ft),

The lunar module, Eagle, orbiting around the Moon after undocking with the command module.

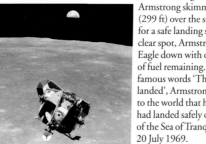

Armstrong and Aldrin realised the computer was guiding them towards a treacherous boulder field. Assuming manual control, Armstrong skimmed Eagle 91 m (299 ft) over the surface, looking for a safe landing site. Seeing a clear spot, Armstrong brought Eagle down with only 20 seconds of fuel remaining. With his now-famous words 'The Eagle has landed', Armstrong announced to the world that he and Aldrin had landed safely on the plains of the Sea of Tranquility, on 20 July 1969.

Buzz Aldrin steps onto the Sea of Tranquility.

Armstrong climbed down the ladder first and stepped on to the surface. Even with the mass of his spacesuit, he only weighed 27 kg (60 lbs) in the weak lunar gravity. Aldrin followed 19 minutes later and the two men collected 22 kg (49 lbs) of rocks and soil and deployed scientific equipment. Among the experiments was a seismometer to monitor any 'moonquakes' and a sail to trap energetic particles from the Sun. The first manned Moon mission was a complete success: Armstrong and Aldrin spent 21 hours 36 minutes on the Moon, and two and a half hours walking on the surface.

Neil Armstrong

American Neil Armstrong (b. 1930) will always be known as the first man on the Moon. Trained as a fighter pilot, Armstrong became an astronaut in 1962 and went on to command Apollo 11, the first lunar landing mission. As he stepped on to the Moon he uttered the unforgettable words: 'That's one small step for man, one giant leap for mankind.'

Apollos 12 and 14

APOLLO 12

THE SECOND MANNED Moon landing started with a jolt. Shortly after lift-off, the Saturn V rocket was struck by lightning, causing a brief moment of panic at mission control and on board. Fortunately, the rest of the flight took place without any trouble and US astronauts Charles Conrad (1930–99) and Alan Bean (b. 1932) touched down in the Ocean of Storms on 24 November 1969, while Richard Gordon (b. 1929) remained in lunar orbit awaiting their return.

Astronaut Charles 'Pete' Conrad inspects Surveyor 3 on the Ocean of Storms.

Apollo 12's primary goal was to find the robotic craft Surveyor 3 which had landed two-and-a-half years earlier, and to bring its camera and other samples back to Earth for analysis. Apollo 12 achieved this perfectly, landing within sight of Surveyor 3.

APOLLO 14

APOLLO 14 WAS commanded by Alan Shepard, America's first man in space. His goal, along with his crew, Edgar Mitchell (b. 1930) and Stuart Roosa (1933–94), was to put the Apollo programme back on track following the near-disastrous Apollo 13 mission. However, the Apollo 14 landing almost failed. The crucial landing radar on board the lunar module Antares refused to work until Shepard and Mitchell were just 72,000 m (22,000 feet) above the surface. Another failed mission would have been a disaster for the Apollo programme, but mission control was able to fix the problem and Antares touched down on 5 February 1971 in the shallow Fra Mauro valley, on the edge of the Mare Imbrium. Antares spent one day, nine-and-a-half hours on the Moon, and Shepard and Mitchell were outside for just over nine hours. They collected samples and set up experiments, including reflectors that would allow lasers shone from Earth to be reflected back and so determine the distance to the Moon with pinpoint accuracy.

Tracks from a hand-pulled equipment trolley stretch back towards the lunar module Antares.

Apollos 15, 16 and 17

APOLLO 15 WAS the first Moon mission to have a heavy emphasis on science. While Alfred Worden (b. 1932) took photographs from orbit, James Irwin (1930–91) and David Scott (b. 1932) landed the lunar module Falcon near the Appenine mountain range on the edge of the Mare Imbrium. Apollo 15 was the first mission to carry a lunar rover, which allowed the astronauts to travel tens of kilometres across the surface and collect rocks from a much wider area than had been possible before. Their most valuable find was a chunk of white rock called anorthosite. At 4.5 billion years old this was one of the oldest samples ever found and would enable scientists to study the Moon's formation. Scott and Irwin were also able to visit the Hadley Rille, a 1.5 km- (0.9 mile-) wide canyon formed by an ancient river of lava.

Apollo 16 was the only mission to visit the lunar highlands. Thomas Mattingly (b. 1936) photographed the surface from

Astronaut James B. Irwin from Apollo 15 surveying Hadley Rille.

orbit in the lunar module, while John Young (b. 1930) and
Charlie Duke (b. 1935) landed in the Descartes highlands on
20 April 1972. With their battery powered lunar rover, the two
men were able to explore several impact craters, including the
1,200 m- (3,940 ft-) wide North Ray Crater, the largest
explored by astronauts on the Moon. During their 71 hours
on the surface, Young and Duke travelled 27 km (17 miles).

It was thought the Descartes highlands were formed by
volcanic activity, and Young and Duke travelled around their
landing site in search of volcanic rocks. But the 97 kg (214 lbs)
of samples they collected during their time on the Moon
contained only ancient rocks, pulverised by aeons
of meteorite bombardment. Apollo 16 proved how hard it was
to understand the planets without visiting them.

*Apollo 16 astronaut Charles M. Duke sets up experiments on the surface
of the lunar highlands.*

The final Apollo Moon mission was also the first to include a fully qualified geologist on the crew. As well as training as an astronaut himself, Harrison Schmitt (b. 1935) had been deeply involved in the geological training of the previous Moon explorers and the scientific selections of landing sites.

Apollo 17 touched down in the Taurus-Littrow valley on the Mare Serenitatis on 11 December 1972. Eugene Cernan (b. 1934) and Schmitt spent 75 hours on the surface, exploring a treasure-trove of geological variety. They collected volcanic samples from the plains as well as from boulders, which had rolled down from the hills. But their greatest find was a patch of orange soil – fallout from a volcanic fire-fountain, which had erupted 3.5 billion years previously.

Harrison Schmitt from Apollo 17 sets off to explore the Taurus-Littrow valley.

THE LEGACY OF APOLLO

THE APOLLO Moon programme spanned four years and represents the first and only human visits to another world. The 12 men who explored the surface returned a total of 382 kg (842 lbs) of rocks and soil samples, as well as a wealth of data from their intense geological surveys. By using this lunar bounty, planetary scientists now believe that the Moon was formed when a planet the size of Mars struck the young Earth 4.5 billion years ago. The core of this intruder was absorbed by our planet, but its rocky exterior was blasted back out into space, where some of the debris coalesced in Earth orbit, eventually forming our only natural satellite.

A treasure trove: samples brought back from the Moon by the Apollo 17 astronauts are studied by scientists.

Salyut

THE FIRST HUMAN OUTPOST

ORBITING THE EARTH at an altitude of around 240 km
(150 miles), space stations are mankind's first outposts in space
and can accommodate astronauts and cosmonauts for months
at a time. The Soviet Union's first space station programme,
Salyut, began in April 1971 and lasted 15 years. In all, there
were seven Salyut stations launched, which accommodated
both civilian and military operations. Each station weighed
over 23 tonnes and their four connected compartments
measured 14.3 m long by 4 m (47 ft long by 13 ft) at their
widest points. The programme had a bad start; after a stay of
three weeks on Salyut 1, cosmonauts Georgi Dobrovolsky
(1928–71), Valdislav Volkov (1935–71) and Viktor Patsayev
(1933–71) suffocated as their re-entry vehicle leaked its
atmosphere into space.

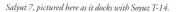

Salyut 7, pictured here as it docks with Soyuz T-14.

Salyut 3 was a military space station and was launched in June 1974. Several days later, cosmonauts Pavel Popovich (b.1930) and Yuri Artyukhin (1930–98) boarded the station from their Soyuz 14 spacecraft and spent a fortnight on board. The second crew assigned to the station were unable to dock and Salyut 3 was incinerated upon re-entry in 1975.

LATER SALYUT STATIONS

THE NEXT Soviet outpost, Salyut 4, was a civilian station which was sent into space in December 1974. The first of the three crews, Aleksei Gubarev (b.1931) and Georgi Grechko (b.1931) spent a month orbiting the Earth. Their activities included making observations of the Sun using the station's solar telescope.

The last military station was Salyut 5, launched in June 1976. It remained in orbit for just over a year – its first crew having to cut their trip short due to a malfunction in the life-support systems.

Salyuts 6 and 7 were upgraded with an extra docking mechanism, allowing unmanned supply craft to dock automatically. Launched in 1982, Salyut 7 operated for four years and finally burned up in the atmosphere in 1991.

Men on a mission: cosmonauts aboard Salyut 6.

Space Stations

SKYLAB

LAUNCHED BY a Saturn V booster on 14 May 1973, Skylab was actually a converted Apollo rocket stage, fitted out for human occupation. The three crews who survived on the station for a total of 171 days lived and worked in comparative luxury. The size of a small house, Skylab contained private sleeping quarters and even a zero-gravity shower. As well as performing hundreds of on-board experiments, the crews also operated the outpost's solar observatory. High above the atmosphere, this collection of television cameras, optical, X-ray and infrared telescopes monitored the Sun constantly for six months and discovered eruptions of billions of tonnes of superheated gases streaming into space.

Skylab was abandoned in 1974, and finally re-entered the Earth's atmosphere and burned up on 11 July 1979.

Skylab provided its inhabitants with plenty of space.

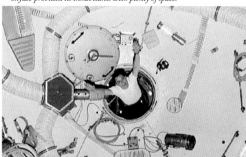

MIR

THE MIR SPACE STATION is the successor to the Salyut
stations of the 1970s and 80s. The first section, the Mir core
module, was launched in February 1986, and contains life-
support systems, living quarters and power systems fed from
external solar panels. Both manned Soyuz capsules and Progress
supply ships dock with ports at either end, and when both are
attached, Mir is approximately 33 m (108 ft) long and 27 m
(88 ft) across. Two Kvant modules were added to Mir in 1987
and 1989, which contain astronomical and biological equipment,
as well as providing extra power and life-support systems. In
1990, the Kristall module was attached to the space station. This
workshop supports experiments on the production of new
materials in weightless conditions, as well as performing extra
biological tests. Kristall is equipped with special docking
machinery designed to allow the experimental Soviet space
shuttle Buran to rendezvous with Mir. Buran was actually
cancelled due to budget cuts and never flew in space with a crew.

Russia's ageing Mir space station.

The last section to be added to Mir was the Priroda module, which docked with the station in April 1996. Priroda's main function is as an Earth observation platform, gathering valuable data on environmental pollution, meteorology and geological resources.

The first crew, Leonid Kizim (b.1941) and Vladimir Solovyev (b.1946), arrived on Mir in March 1986. They activated the station and unloaded cargo and supplies from the two Progress vehicles, which had previously docked automatically. They spent two months on board before undocking their Soyuz capsule and heading for the redundant Salyut 7 station, which was still in orbit. They spent 50 days on Salyut 7 before transferring equipment back to Mir.

Mir cosmonauts undertake some essential maintenance.

There have been over 30 missions to Mir, consisting of crews of two or three people. Astronauts from several European and non-European countries, including Britain, have flown on the station.

INTERNATIONAL SPACE STATION

THE INTERNATIONAL space station (ISS) is a collaborative effort of 16 countries around the world, although the USA and Russia are providing most of the actual hardware. Between 1995 and 1998 the space shuttle docked nine times with Mir, transferring crews and allowing the Russian and US space agencies to develop their cooperation in space, essential to the future of ISS.

The first section of ISS, the Russian-built Zarya module, was launched in November 1998, and the second module, Unity, was attached one month later. Assembly will be complete in the next few years, and the final station will be the size of a jumbo jet, with a mass of 450 tonnes.

Powered by huge solar panels, ISS will have crew living in space, as well as seven research laboratories for conducting scientific experiments and observations. A Russian Soyuz craft will be permanently docked, acting as an emergency escape vehicle while a 17 m- (56 ft-) long robot arm will be able to manipulate cargo outside the station.

An artist's impression of how the International Space Station will look when it is completed.

The Space Shuttle

EVEN WHILE THE APOLLO missions were exploring the Moon, NASA was devising a re-usable space shuttle. The Mercury, Gemini and Apollo spacecraft could only be used once and the US space agency hoped that a re-usable vehicle would drastically cut the huge cost of sending astronauts into space. The US government approved the plans in 1972 and NASA began designing the space shuttle in earnest.

THE FIRST RE-USABLE SPACECRAFT

THIS SPACECRAFT consists of several components: a delta-winged orbiter, a huge external fuel tank, and two rocket boosters. The orbiter has a wingspan of 24 m (79 ft), which allows it to glide back to Earth after completing a mission. The underside is entirely covered with black insulating tiles that protect the shuttle and the crew from the intense heat caused by the friction of re-entering Earth's atmosphere. Behind the crew cabin at the front, an 18.3 m- (60 ft-) long cargo bay can open in space to launch and retrieve satellites. Fuel for the shuttle's own engines is supplied by the external tank, which is jettisoned once empty to burn up in the atmosphere. The rocket boosters provide extra thrust and are also jettisoned when used up, but they parachute back to Earth to be used again.

Launched on 12 April 1981, the first space shuttle, named Columbia, was piloted by Americans Robert Crippen (b. 1937) and John Young, a veteran of the Gemini and Apollo missions. The new spacecraft performed excellently and Young and Crippen completed 36 orbits of the Earth and tested the orbiter's performance and manoeuvrability. Fifty-four hours after launch, Columbia touched down safely at Edward's Airforce Base in California. This was the first time that a spacecraft had landed on Earth using wheels.

Space shuttle Columbia blasts off on its first mission in 1981.

...cCandless makes the first untethered spacewalk in an MMU.

Milestones and Achievements

SINCE THE LAUNCH of Columbia in 1981, 100 shuttle missions have been launched. NASA currently has three shuttles: Discovery, Atlantis and Endeavour.

...itoring and medical resear...

Magellan being launched on its mission to Venus in 1989.

THE RISKS OF SPACE TRAVEL

THE SPACE SHUTTLE Challenger exploded 73 seconds after launch in January 1986, killing its crew of seven men and women. NASA's space shuttle missions were grounded for more than two-and-a-half years while an investigation, led by Neil Armstrong, was undertaken. This discovered that the explosion had been caused by flames leaking from one of the booster rockets and burning through the huge liquid-fuel tank, creating a catastrophic fireball. More recently, in February 2003, the Space Shuttle Columbia broke up on re-entry. All seven crew members died. An investigation concluded that a piece of foam from the fuel tank hit the wing on launch, fatally damaging it. These tragedies reminded the world of the difficulty and danger of human space exploration.

SATELLITES

THE SHUTTLE'S cargo bay is highly versatile, and has been used to send numerous satellites into orbit around the Earth. Additionally, the shuttle has been used to launch robotic missions to other planets: in 1989, both the Magellan mission to Venus, and the Galileo mission to Jupiter were carried into orbit by Atlantis. The following year, Discovery was launched carrying the Ulysses probe, which was to explore the poles of the Sun.

SPACELAB

ON SOME 23 spaceflights, the shuttle has been fitted out with an advanced research centre, called Spacelab. This scientific laboratory was first taken into space on board Columbia in 1983, and performed 73 different experiments, including astronomy, Earth environmental mon_____.

SPACEWALKS AND REPAIRS

ONE OF THE best-remembered images from the space shuttles' flights is that of astronaut Bruce McCandless (b. 1937). In 1984 McCandless tested the Manned Manoeuvring Unit (MMU), and achieved the first untethered spacewalk. The MMU is a self-contained spacecraft which an astronaut straps on. The astronaut is then able to move around space using the MMU's own thrusters.

Bruce M

Perhaps the most famous shuttle missions have been the three servicing missions to the Hubble Space Telescope. The shuttle used its jointed robotic arm to dock with the telescope, holding it in place while astronauts tackled the intricate job of removing and replacing old components with new ones.

The longest shuttle flight so far was that of Columbia, which orbited the Earth for almost 17 days in 1996. This flight carried the Life and Microgravity Spacelab, which contained over 40 experiments, operated remotely by scientists based in Europe and the USA.

The shuttle docking with Mir, forming the largest-ever structure in space.

Although the space shuttle has failed to live up to its expectations of a 'cheap' alternative to expendable spacecraft, it has enabled NASA to keep sending astronauts into space. This has in turn allowed shuttle crews to gain crucial experience needed for the ISS, currently under construction.

Important practise for shuttle crews have been the joint operations with the Russian space agency. Between June 1995 and June 1998, the shuttle made nine flights that docked with the Russian Mir space station, forming the largest-ever structure in orbit.

SURVIVING IN SPACE

Introduction

M ANKIND'S EFFORTS to explore space are both high-profile and expensive. Some argue that exploration comes naturally to our inquisitive species, while others claim that the enormous cost should be diverted to problems on Earth.

SPACE PRIORITIES

SPACE IS DEADLY to life. Everything required by humans to live must be hauled into space with them, and this increases the price tag dramatically. Some scientists argue against human spaceflight on the grounds that it drains resources away from unmanned planetary probes, which have yielded so much knowledge. The space shuttle was developed with the promise of cheaper human access to space, but has proven more expensive than the expendable rockets it was to replace. The Russian manned space programme is foundering from lack of funds. With the end of the space race more than 30 years ago, NASA needs to decide where it is going in space – and why. There are tentative plans for human missions to Mars, which would cost hundreds of billions of dollars, but no deadline has been set.

POTENTIAL BENEFITS

ONE CANNOT ALWAYS predict the benefits of space exploration. When the crew of Apollo 8 photographed the whole Earth set against the foreground of the desolate lunar horizon, the image left a significant impression on millions of people. For the first time, the Earth looked small and fragile, with limited resources and a fragile ecosphere, and this image provided a boost to the environmental movement. Who knows what is to be gained from expanding our civilisation into the Universe? Perhaps only time will tell.

Is it better to have more unmanned (cheaper) missions, like the Rosetta mission (launched in 2004) than manned (expensive) ones?

Problems of Life in Space

WEIGHTLESSNESS

HUMAN BEINGS evolved on Earth, a planet with gravity. The weightless environment of space is completely unnatural. Before the first astronauts left the Earth, both the Soviets and the Americans sent animals into orbit, to discover whether or not living things could actually function at all in weightless conditions.

Upon reaching orbit, astronauts can often become disorientated, with no up or down to rely on, and more than one space traveller has become nauseous.

EATING IN SPACE

EATING IN SPACE can be a tricky business. Food floating around the cabin and fouling up the instruments would be unthinkable, so astronauts take food that tends to stick to a utensil and not drift off. In the early days of spaceflight this severely restricted an astronaut's diet. But today's astronauts and cosmonauts have a choice of hundreds of different meals, often held together by a sauce. Drinks are supplied in sealed containers and are drunk using a straw.

The problems of eating in a weightless environment have now been overcome.

MUSCLE LOSS

DURING spaceflight the muscles in the human body begin to deteriorate, as in weightless conditions legs do not have to walk and arms are not required to lift things up, and the merest touch can propel an astronaut across a room. To prevent this, on long-duration flights, Mir cosmonauts need to use an exercise bicycle or a treadmill for two-and-a-half hours each day.

If humans are ever going to make a trip of several months to Mars, they will have to be in top physical condition when they arrive. How to keep fit in weightless conditions is one of the major studies of today's manned space programmes.

How to keep fit whilst in zero gravity.

Spacesuits

THE FIRST SPACESUITS were modified versions of high-altitude jet-fighter pressure suits. They were originally designed for spaceflights on which the astronaut would stay inside the spacecraft for the whole mission. As the human exploration of space has become more complex and ambitious, spacesuits have evolved to become increasingly technically advanced. On modern space missions, astronauts and

cosmonauts are able to wear normal casual clothing for most of the flight, with spacesuits required only for launch, spacewalks and re-entry

A MINI ATMOSPHERE

SPACESUITS ARE designed to be a substitute for the Earth's atmosphere. A mini environment, they must provide a self-contained, independent life-support system, which cannot afford to malfunction. As well as providing oxygen for breathing and communications, the suit must be able to regulate the

By today's standards, Alan Shepard's 1961 suit seems very primitive.

An astronaut being prepared for space-simulation in a swimming pool.

temperature of the astronaut inside – in the near-vacuum of space, the temperature can reach 120°C (248°F) in the sunlight and -150°C (-238°F) in the shade.

The spacesuits used by the astronauts during the Apollo missions had to be exceptionally durable and versatile. As well as providing enough life support to last up to seven hours at a time on the lunar surface, they had to protect the wearer from jagged rocks and other lunar hazards, such as the very fine dust.

One of the problems of a pressurised suit is that the astronaut is essentially inside a balloon. Special mechanisms at the joints of the suit have to be used so that the astronaut can bend the knees and elbows without too much physical exertion. Astronauts have also had to train extensively to be able to work complex and intricate tools through the thick gloves.

Today's spacesuits are the culmination of decades of research and development. The early spacesuits were custom-made for individual wearers, but nowadays space shuttle astronauts use suits which have many interchangeable parts of different sizes, to fit different people.

Repairs in Space

PLANNED REPAIRS

WHEN NASA'S SKYLAB space station was launched in May 1973, the orbital outpost ran into problems. A malfunction had left a critical thermal shield in ribbons, allowing heat from the Sun to bake the inside of Skylab. Also, one of the solar panels had become jammed and failed to open. When the first crew arrived 11 days later, their first task was to erect a new shield, to reflect the sunlight and make the station habitable. A fortnight later, two of the astronauts performed a three-and-a-half hour spacewalk, in which they were able to free the damaged solar panel and restore full power to Skylab.

HUBBLE TROUBLE

THE DELICATE PRECISION required while working in space was demonstrated during the first shuttle mission to repair the Hubble Space Telescope in 1993. With the telescope upright in the shuttle's cargo bay, American astronaut Story Musgrave (b. 1935) perched on the end of the robotic arm and had to manipulate tiny screws using his bulky gloves, to gain access to the internal wiring of the satellite observatory.

EMERGENCY REPAIRS

SOME REPAIRS THOUGH, have been on-the-spot jobs in response to a crisis. On Apollo 13, while the crew of three sheltered in the lunar module Aquarius, the air filters that removed the carbon dioxide from the astronauts' breath were becoming saturated. The filters on the crippled command

Essential repair work being carried out on Hubble in 1993.

module would not fit those on Aquarius, so the crew hastily built an adaptor using, among other items, plastic bags, the cover of the flight manual, and one of the astronaut's socks!

Close Shaves

FROM THE PIONEERING FLIGHTS of the early 1960s to today's space shuttle missions, sending humans into space has never been routine or easy. Almost every spaceflight has had minor problems, which the crew have had to overcome. However, some missions have come perilously and famously close to disaster.

HOUSTON, WE'VE HAD A PROBLEM

ON 13 APRIL 1970, 56 hours into the mission and more than half way to the Moon, Apollo 13 suffered a crippling explosion. One of the command module Odyssey's oxygen tanks ruptured, leaving US astronauts James Lovell, Fred Haise

As a result of an explosion 56 hours into the mission, an external panel came away from Apollo 13's service model.

b. 1933) and Jack Swigert (1931–82) critically short of water, electrical power and breathing oxygen. For the next four days the whole world held its breath as the three men, sheltering inside the lunar module Aquarius, having abandoned Odyssey, passed around the far side of the Moon and headed back to Earth. In what has to have been NASA's finest hour, the team at mission control in Houston guided the dehydrated and exhausted crew to a safe splashdown in the Pacific Ocean on 17 April 1970.

MIR MORTALS

MORE RECENTLY, on 23 February 1997, tragedy was averted when the crew of Mir just managed to extinguish a fire on board the ageing Russian space station. Then, just four months later, an unmanned Progress supply craft collided with the station, rupturing the hull and crippling the main computer. This potentially fatal situation left the crew unable to activate the escape pod. As if this was not enough, the main computer completely failed again in August 1997, leaving the space station tumbling temporarily out of control. Mir is still operational today, having been in space 10 years longer than planned.

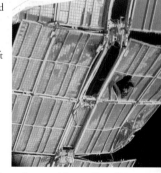

Damage to Mir caused by a collision with a supply craft.

Space Debris

E VER SINCE THE first rocket launches into space, we have been polluting the environment above the Earth. Pieces of junk and debris from spacecraft have been accumulating in orbit around the Earth and are beginning to present a distinct hazard to both manned spacecraft and artificial satellites. The space junk ranges in size from dead satellites and spent rocket stages to individual flecks of paint from spacecraft.

Satellite deployment releases debris into outer space.

The hazard posed by this junk is due to the very high speeds at which it orbits Earth: at upwards of 27,000 kph (17,000 mph), a 1-cm (½ in) fragment can easily cripple a spacecraft, and a 1-mm piece can still cause serious damage.

AN ORBITAL MINEFIELD

IN FEBRUARY 1997, while the space shuttle Discovery was docked with the Hubble Space Telescope, mission control in Houston ordered the crew to fire the shuttle's manoeuvering jets in order to avoid a lethal chunk of debris. This piece of space junk was from a Pegasus rocket, which had exploded a few years earlier. Hubble was launched into space in April 1990,

and in that time it has been pitted with hundreds of holes
ranging in size from less than 1 mm to a few centimetres. In
1996, a small British-built satellite called CERISE was disabled
when it was struck by a fragment from an Ariane rocket, which
had blown up shortly after launch several years earlier.

Surveys carried out in recent years by NASA estimate that
there are perhaps 100,000 pieces of debris 1 cm (½in) or more
across and 1,000,000 pieces 1 mm across in orbit around
the Earth.

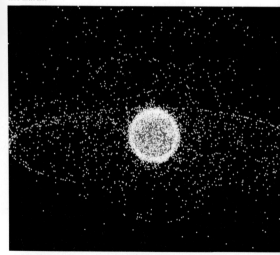

The distribution of space debris in Earth's orbit.

THE FUTURE OF SPACE EXPLORATION

Introduction

THE HUMAN RACE has barely started to explore space. Since the Apollo programme, no human has returned to the Moon and astronauts have not ventured beyond the Earth's orbit. But space exploration continues in earnest. Astronomers may be only years away from finding the tell-tale signs of extraterrestrial life, and from discovering the eventual fate of the whole Universe.

Robot probes continue to explore the planets. Spacecraft are currently mapping Mars and, for the very first time, orbiting an asteroid. The Cassini-Huygens mission arrived at Saturn in 2004 and begin an intensive survey of the ringed planet from orbit. The Huygens probe landed on the surface of the cloudy moon Titan, suspected to have oceans of methane and complex chemistry similar to Earth's before life began.

The completion of the ISS in the next few years will be seen as mankind's latest stepping stone into the Universe, showing the

A NASA computer-generated image showing the Hugyens lander on Titan's surface in 2004.

way to the next great endeavour in space exploration: a human mission to Mars. Perhaps within the next 20 years, a spacecraft will depart from an orbiting space station, carrying a crew of engineers and scientists to the red planet. This journey will take at least several months and could involve a stay on the surface lasting a few more. NASA is currently testing a prototype Martian base on Earth in an attempt to perfect the art of recycling water and oxygen. Eventually, during the twenty-first century, permanent bases on the Moon and Mars will sustain human colonies, and perhaps even tourist resorts.

Mars is next on the list for human exploration.

Giovanni Domenico Cassini

Cassini (1625–1712) was invited from Bologna University, Italy, where he had been Professor of Astronomy for 21 years, to become the first director of the newly built Paris Observatory in 1671. He discovered four of Saturn's moons (Iapetus, Rhea, Tethys and Dione) and in 1675 discovered the gap in Saturn's ring system, which was named the Cassini Division.

Science Fiction

SCIENCE FICTION has had a huge impact on actual science. It is possible that, without the imaginations of authors such as the Frenchman Jules Verne (1828–1905), the real scientific advancements would have taken longer to come about.

One of the pioneering science fiction stories was *From the Earth to the Moon*, published in 1865. The author, Jules Verne used the scientific thinking of the day to send a group of travellers to the Moon. The Apollo missions 100 years later drew many parallels with the fictional voyage. It is possible, however, that the similarities were self-fulfilling rather than coincidence. Many of the scientists and engineers who developed spaceflight had read Verne's book, and astronauts such as Neil Armstrong and Yuri Gagarin had been inspired by its lunar voyage as children.

TECHNOLOGY AT WARP SPEED

Captain Kirk and his crew aboard the starship Enterprise.

STAR TREK is probably the most famous science fiction creation. Boldly going since the late 1960s, the crew of the starship Enterprise has entertained millions of TV viewers and movie-goers with their super-advanced technology. But with actual technology catching up, the writers of *Star Trek* are finding it more difficult to devise new futuristic gadgets for the crew. The infamous flip-open communicator used by Captain Kirk can now be seen in the homes and pockets of millions of people, as mobile phones! Similarly, scientists are testing a real version of Enterprise medic Dr McCoy's 'hypo-spray', using a fine jet to administer medication without a needle.

Because of the enormous distances between the stars, spacecraft would require many years to travel between them. Although the laws of physics forbid anything travelling faster than the speed of light (300,000 m/sec / 984,000 ft/sec), the 'warp drives' in Star Trek are designed to do just that. However, the cosmologist Stephen Hawking (b. 1942) has stated, half jokingly, that he is working on the theory of 'faster-than-light' propulsion, which could allow humans to cross the gulf between the stars and explore the Universe.

Stephen Hawking

Stephen Hawking was born in 1942, 300 years after the death of Galileo. His position of Lucasian Professor of Mathematics at Cambridge University, was once held by Isaac Newton. One of his most important discoveries is that black holes emit radiation, and will slowly evaporate. His book, *A Brief History of Time*, has been translated into over 30 languages.

Space Tourism

EVER SINCE the first human spaceflights, millions of ordinary people have been waiting and hoping for their chance to orbit the Earth or to travel beyond to the Moon and the planets. In the 1960s experts were confident that, by the end of the twentieth century, space tourism would be a thriving industry. Space planes would be carrying passengers around the world in 90 minutes, and orbital hotels would be the most glamorous tourist destinations. But all this has yet to happen. All space launches are still controlled by governments, and little of their budgets is devoted to developing space tourism. Only when private individuals and companies can exploit space technology will space travel become commonplace.

THE NEW SPACE RACE

SOME OF THE world's largest construction companies already have concept designs for space hotels and bases on the Moon and Mars, but the investment required would run to many billions of pounds. In the meantime, within the next couple of decades it is likely that short sub-orbital trips, similar to Alan Shepard's 15-minute flight of 1961, will be available to the super-rich. And a new company called Mircorp has leased the veteran Mir space station from the Russian government, with plans to use it for commercial purposes. Also, many amateur groups are building rockets that may be able to launch satellites within the next few years.

But access to space for the many who are not multi-millionaires will have to wait. As technology continues to advance, there will certainly be some people holding their breath for the first genuine tickets to space to become available.

An artist's conception of how a lunar settlement might be inhabited by humans.

Wormholes and Time Travel

HYPERSPACE CONDUITS

WORMHOLES AND time travel are usually confined to the realms of science fiction, but some scientists are seriously investigating the physical theory behind these bizarre concepts. Wormholes, if they exist, are conduits connecting one region of space to another. A space traveller might be able to pass through the conduit instantaneously, in effect, travelling much faster than the speed of light. Alternatively, once you enter a wormhole, perhaps via a black hole, you could emerge in a totally different universe, or perhaps you would travel in time.

These strange phenomena were first predicted by some as a solution to Albert Einstein's (1879–1955) general theory of relativity. Although science has been unable to show that they are impossible, as yet there is no evidence that wormholes do exist.

Like in science fiction films, time travel could, one day, become a reality.

THE GRANDPARENT PARADOX

THE MAIN PROBLEM posed by time travel is a scenario known as 'the grandparent paradox'. If you could travel backwards through time, then you could kill your grandparents and prevent yourself from ever being born in the first place! How then could you travel back to the future? Another problem is that if time travel were possible, then why are we not inundated with travellers from the future? One possible solution involves infinite parallel universes. If you did travel back in time and meet your grandparents, you could spark off a new, parallel universe in which you were never born. Your original universe would still exist, unaffected by the new universe. Ultimately, while wormholes and time travel might exist on paper, it is possible they will remain there forever.

Albert Einstein

German-born physicist Albert Einstein (1879–1955) was one of the greatest scientific thinkers of the twentieth century. He is best known for his equation $E=mc^2$, the calculation for his special theory of relativity published in 1905, and for his general theory of relativity of 1916. These theories had a huge impact in the field of physics, in particular in explaining the Sun and the speed of light.

Spacecraft of the Future

THE X-33

THE X-33 IS CURRENTLY being developed jointly by NASA and the commercial aerospace industry. This new concept for a successor to the space shuttle is designed to be launched without the need for extra expendable rocket stages or strap-on booster rockets. Using a revolutionary engine design called an 'aerospike', the unmanned X-33 will travel at 19,000 kph (11,800 mph) to an altitude of almost 90 km (56 miles). If successful, the prototype will be used to construct a larger version, the VentureStar, capable of carrying payloads and people into orbit.

NEW PROPULSION TECHNIQUES

BEYOND THE VentureStar, experiments are underway to test the engines for future spacecraft. NASA's Deep Space 1 mission was launched in October 1999. Its main goal has been to test a new system of propulsion. Instead of burning rocket fuel and oxygen to provide thrust, Deep Space 1 uses solar electric propulsion. Around 10 times more efficient than conventional thrusters, this 'ion drive' electrically charges atoms of xenon in the engine. A magnetic field then accelerates the charged atoms, or ions, out of the spacecraft as high-velocity exhaust, providing the propellant.

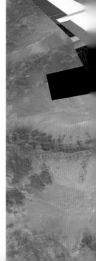

As ion drives evolve, future spacecraft will become lighter and more efficient, slashing the cost of potential human missions to the planets. But if we are ever going to cross the vast distances between the stars, and colonise the Galaxy, humanity must develop spacecraft capable of faster-than-light travel. The current laws of physics forbid this, but it is possible in the future that a solution may arise from the study of black holes and the fabric of space and time.

The future of space travel?: A computer-generated image of the VentureStar.

COMPENDIUM

How to Start Stargazing

THE BEST OPTICS with which to start stargazing are your own eyes. The other alternative is a telescope, but a good telescope can cost hundreds of pounds. Some cheap telescopes are available, but tend to be over-powered for their size. With too high a magnification, the image in the eyepiece of a small telescope can be disappointingly dim. How much you spend depends on how seriously you are going to take up stargazing. Location is paramount whatever method you choose – city lights and pollution drown out all but the brightest stars.

Light pollution from cities can drown out most of the stars in the night sky.

A CELESTIAL FEAST, VISIBLE TO THE NAKED EYE

CHOSE A CLEAR, moonless night. Once away from the city
lights, a dark location, a good view of the sky, and perhaps a
deck chair are all that are needed. Good star maps are readily
available and can be surprisingly easy to use after a little
practise. The recognisable patterns of Ursa Major and Orion act
as useful navigation points to the other constellations. If the sky
is clear, the great band of the Milky Way can be seen arcing
across the sky. This diffuse light comes from billions of distant
stars along the plane of our Galaxy. Shooting stars are common,
especially in the middle of August and November, when one or
two per minute can be seen. With a good pair of binoculars, the
moons of Jupiter are visible, changing positions from night to
night. But for beginners, the best place to start is by simply
getting used to the constellations, gazing at the myriad of stars
and occasionally
slowly sweeping the
sky with binoculars.
Once you know
your way around the
sky, a standard pair
of 7 x 50 unit
binoculars will
reveal a vast visual
feast of bright
nebulae, star
clusters and even the
Andromeda Galaxy.

*The sky at night,
with the Milky Way
clearly visible.*

Useful Data

THE PLANETS

The information below is arranged in the following order of planets:

	MERCURY	VENUS	EARTH	MARS	JUPITER	SATURN	URANUS	NEPTUNE	PLUTO
Diameter at equator kilometres (miles)	4,878 (3,031)	12,103 (7,521)	12,756 (7,927)	6,786 (4,217)	142,980 (88,848)	120,540 (74,904)	51,120 (31,766)	49,530 (30,778)	2,280 (1,417)
Diameter relative to Earth (Earth = 1)	0.38	0.95	1.00	0.53	11.0	9.41	4.11	3.96	0.18
Mass relative to Earth (Earth = 1)	0.055	0.815	1.00	0.107	318	95	14.54	17.15	0.002
Average density relative to water	5.43	5.2	5.5	3.9	1.32	0.69	1.32	1.64	2.1
Volume relative to Earth (Earth = 1)	0.06	0.88	1.00	0.15	1,316	755	67	57	0.015
Average temperature in degrees Celsius (Fahrenheit) (S-surface, C-clouds)	350* (662) (S)	-33* (-27) (C)	22 (72) (S)	-23 (-9) (S)	-150 (-238) (C)	-180 (-212) (C)	-210 (-346) (C)	-220 (-364) (C)	-230 (-382) (S)

* day   **Mercury night -170 (-274) (S)   Venus night 480 (896) (S)

Surface gravity relative to Earth									
0.38	0.9	1.0	0.38	2.64	1.16	1.11	1.21	0.06	
Atmosphere (main components)									
None	Carbon dioxide	Nitrogen, Oxygen	Carbon dioxide, Argon	Hydrogen, Helium	Hydrogen, Helium	Hydrogen, Helium, Methane	Hydrogen, Helium, Methane	Hydrogen, Helium, Methane	Nitrogen
Tilt of axis (degrees)									
0.0	177.4	23.27	24.46	3.05	26.44	97.53	28.48	119.6	
Inclination of orbit to ecliptic (degrees)									
7	3.4	0	1.9	1.3	2.5	0.8	1.8	17.2	
Period of revolution around the Sun									
88 days	225 days	365 days	687 days	11.9 years	29.5 years	84 years	165 years	248 years	
Period of rotation on axis (in days) relative to the stars									
Actual length of day (sunrise to sunset) given in brackets if it is much different									
59 days (176 days)	243 days (117 days)	23 hrs 56 mins (24 hours)	24 hrs 37 mins	9 hrs 55 mins	10 hrs 39 mins	17 hrs 14 mins	16 hours 7 mins	6 days 9 hours	
Mean distance from Sun millions of km (millions of miles)									
57.9 (35.9)	108.2 (67.2)	149.6 (92.9)	227.9 (141.6)	778.3 (483.6)	1,427 (887)	2,869.6 (1,783.1)	4,496.6 (2,794.1)	5,900 (3,666)	
Mean orbital velocity km per second (miles per second)									
47.9 (29.7)	35 (22)	29.8 (18.5)	24.1 (14.9)	13.1 (8.1)	9.7 (6.0)	6.8 (4.2)	5.4 (3.3)	4.7 (2.9)	

MAJOR MOONS OF THE PLANETS

The information below is arranged in the following order:

MEAN DISTANCE FROM CENTRE OF PLANET KM (MILES)	ORBITAL PERIOD (DAYS)	DIAMETER KM (MILES)
Earth		
Moon 384,400 (238,886)	27.321	3,475.6 (2,159.7)
Mars		
Phobos 9,270 (5,760)	0.32	17 × 14 × 11 (11 × 9 × 7)
Deimos 23,400 (14,541)	1.26	9 × 7 × 6 (6 × 4 × 4)
Jupiter		
Metis 127,960 (79,514)	0.295	40 (25)
Adrastea 128,980 (80,770)	0.298	26 × 20 × 16 (16 × 12 × 10)
Amalthea 181,300 (112,660)	0.498	262 × 146 × 143 (163 × 91 × 89)
Thebe 221,900 (137,889)	0.675	68 × 56 (42 × 35)
Io 421,600 (261,982)	1.769	3,637 (2,260)
Europa 670,900 (416,897)	3.551	3,130 (1,945)
Ganymede 1,070,000 (664,898)	7.155	5,268 (3,274)
Callisto 1,880,000 (1,168,232)	16.689	4,806 (2,986)
Leda 11,094,000 (6,893,812)	238.7	10 (6)
Himalia 11,480,000 (7,133,672)	250.6	170 (106)
Lysithea 11,720,000 (7,282,808)	259.2	24 (15)
Elara 11,737,000 (7,293,372)	259.7	80 (50)
Ananke 21,200,000 (13,173,680)	631 (retrograde*)	20 (12)
Carme 22,600,000 (14,043,640)	692 (retrograde)	30 (19)
Pasiphae 23,500,000 (14,602,900)	735 (retrograde)	36 (22)
Sinope 23,700,000 (14,727,180)	758 (retrograde)	28 (17)

*The moon orbits the planet in the opposite direction to the planet's spin

Saturn

Pan 133,600 (83,019)	0.57	12 (8)
Atlas 137,670 (85,548)	0.602	37 × 34 × 27 (23 × 21 × 17)
Prometheus 139,350 (86,592)	0.613	148 × 100 × 68 (92 × 62 × 42)
Pandora 141,700 (88,052)	0.629	110 × 88 × 62 (68 × 55 × 39)
Epimetheus 151,420 (94,092)	0.694	194 × 190 × 154 (121 × 118 × 96)
Janus 151,470 (94,124)	0.695	138 × 110 × 110 (86 × 68 × 68)
Mimas 185,540 (115,295)	0.942	421 × 395 × 385 (262 × 245 × 239)
Enceladus 238,040 (147,918)	1.370	512 × 495 × 488 (318 × 308 × 303)
Tethys 294,670 (183,108)	1.888	1,046 (650)
Telesto 294,670 (183,108)	1.888	30 × 25 × 15 (19 × 16 × 9)
Calypso 294,670 (183,108)	1.888	30 × 16 × 16 (19 × 10 × 10)
Helene 377,410 (234,523)	2.737	35 (22)
Dione 377,420 (234,529)	2.737	1,120 (696)
Rhea 527,040 (327,503)	4.518	1,528 (950)
Titan 1,221,860 (759,264)	15.945	5,150 (3,200)
Hyperion 1,481,100 (920,356)	21.277	360 × 280 × 225 (224 × 174 × 140)
Iapetus 3,651,300 (2,268,918)	79.331	1,436 (892)
Phoebe 12,954,000 (8,049,616)	550.4 (retrograde)	230 × 220 × 210 (143 × 137 × 130)

Uranus

Cordelia 49,471 (30,741)	0.330	26 (16)
Ophelia 53,796 (33,429)	0.372	30 (19)
Bianca 59,173 (36,770)	0.433	42 (26)
Cressida 61,777 (38,388)	0.463	62 (39)
Desdemona 62,676 (38,947)	0.475	54 (34)
Juliet 64,372 (40,001)	0.493	84 (52)
Portia 66,085 (41,065)	0.513	108 (67)
Rosalind 69,941 (43,461)	0.558	54 (34)
Belinda 75,258 (46,765)	0.622	66 (41)
1986U10 75,000 (46,605)	0.62	40 (25)

MEAN DISTANCE FROM CENTRE OF PLANET KM (MILES)	ORBITAL PERIOD (DAYS)	DIAMETER KM (MILES)
Uranus (continued)		
Puck 86,000 (53,440)	0.762	154 (96)
Miranda 129,400 (80,409)	1.414	472 (293)
Ariel 191,000 (118,687)	2.520	1,158 (720)
Umbriel 266,300 (165,479)	4.144	1,169 (726)
Titania 435,000 (270,309)	8.706	1,578 (981)
Oberon 583,500 (362,587)	13.463	1,523 (946)
Caliban 7,100,000 (4,412,000)	579.4	60 (37)
1999U1 10,000,000 (6,000,000)	?	40 (25)
Sycorax 12,200,000 (7,581,000)	1284	120 (75)
1999U2 25,000,000 (16,000,000)	?	30 (19)
1999U3 ?	?	40 (25)
Neptune		
Naiad 48,000 (29,827)	0.296	54 (34)
Thalassa 50,000 (31,070)	0.312	80 (50)
Despina 52,500 (32,624)	0.333	180 (112)
Galatea 62,000 (38,527)	0.429	150 (93)
Larissa 73,600 (45,735)	0.554	192 (119)
Proteus 117,600 (73,077)	1.121	416 (259)
Triton 354,800 (220,473)	5.877 (retrograde)	2,705 (1,681)
Nereid 1,345,500–9,688,500 (836,094–6,020,434)	360.16	240 (149)
Pluto		
Charon 19,640 (12,204)	6.387	1,212 (753)

Spacecraft Chronology

MISSIONS TO THE MOON & PLANETS

MISSIONS TO THE MOON

USSR

- Luna and Zond missions. 28 probes launched between 1959 and 1976. The missions were mainly successful, photographing the Moon's far side, orbiting and landing, with four returning soil samples. Lunas 17 and 21 carried remote-controlled rovers in 1970 and 1973 – both were very successful.

USA

- Ranger missions. nine probes launched between 1961 and 1965. These missions were designed to plunge to the lunar surface, taking images until the last second before impact. Only the last three were fully successful.
- Orbiter missions. Five probes launched between 1966 and 1967. Between them they returned 1,474 detailed pictures of the surface from orbit, searching for Apollo landing sites.
- Surveyor missions. Seven probes launched between 1966 and 1968. Designed to land softly on the surface. Five missions succeeded, photographing the landscape and analysing soil.

Recent US missions

- Clementine launched 24 January 1994. Mapped the whole Moon in different wavelengths of light.
- Lunar Prospector launched 7 January 1998. Mapped the Moon's gravitational field.

Apollo

- Apollo 8, 21–27 December 1968. Crew: Frank Borman, James Lovell, William Anders. First manned mission to orbit the Moon.
- Apollo 10, 18–26 May 1969. Crew: Thomas Stafford, John Young, Eugene Cernan. Orbited the Moon, practised landing techniques.
- Apollo 11, 16–24 July 1969. Crew: Neil Armstrong, Michael Collins, Edwin 'Buzz' Aldrin. First manned lunar landing on 21 July in the Sea of Tranquility.
- Apollo 12, 14–24 November 1969. Crew: Charles Conrad, Richard

Gordon, Alan Bean. Landed in the Ocean of Storms, close to the earlier Surveyor 3 unmanned probe.
- Apollo 13, 11–17 April 1970. Crew: James Lovell, John Swigert, Fred Haise. Crew almost killed by on-board explosion. Moon landing cancelled and crew survived perilous trip back to Earth.
- Apollo 14, 31 January–9 February 1971. Crew: Alan Shepard, Stuart Roosa, Edgar Mitchell. Landed in Fra Mauro highlands.
- Apollo 15, 26 July–7 August 1971. Crew: David Scott, Alfred Worden, James Irwin.
- Landed near Apennine mountain range. First mission to include a lunar rover.
- Apollo 16, 16–27 April 1972. Crew: John Young, Thomas Mattingly, Charles Duke. Landed in the Descartes highlands, found no evidence of predicted volcanic rocks. Used second lunar rover.
- Apollo 17, 7–19 December 1972. Crew: Eugene Cernan, Ronald Evans, Harrison Schmitt. Last manned Moon mission. Geological survey of the spectacular Taurus-Littrow valley. Used third lunar rover.

Japan
- Hiten-Hagoromo launched 24 Jan 1990. First Japanese planetary mission – flyby and orbiter.

MISSIONS TO MERCURY
- Mariner 10 (USA) launched 3 November 1973 arrived 29 March 1974. Encountered planet three times in 1974 and 1975.

MISSIONS TO VENUS
- Venera 1 (USSR) launched 12 February 1961. Failed 7.5 million km (47 million miles) from Earth.
- Mariner 1(USSR) launched 22 July 1962. Launch failed.
- Mariner 2 (USA) launched 27 August 1962 arrived 14 December 1962. Successful flyby.
- Zond 1 (USSR) launched 2 April 1964. Failure due to loss of contact shortly after launch.
- Venera 2 (USSR) launched 12 November 1965 arrived 27 February 1966. Successful flyby.
- Venera 3 (USSR) launched 16 November 1965 arrived 1 March 1966. Destroyed during landing attempt.
- Venera 4 (USSR) launched 12 June 1967 arrived 18 October 1967.

Data transmitted during landing attempt.
- Mariner 5 (USA) launched 14 June 1967 arrived 19 October 1967. Successful flyby.
- Venera 5 (USSR) launched 5 January 1969 arrived 16 May 1969. Destroyed during landing attempt.
- Venera 6 (USSR) launched 10 January 1969 arrived 17 May 1969. Destroyed during landing attempt.
- Venera 7 (USSR) launched 17 August 1970 arrived 15 December 1970. Survived 23 minutes after landing.
- Venera 8 (USSR) launched 26 March 1972 arrived 22 July 1972. Survived 50 minutes after landing.
- Venera 9 (USSR) launched 8 June 1975 arrived 21 October 1975. Transmitted one picture of the surface.
- Venera 10 (USSR) launched 14 June 1975 arrived 25 October 1975. Transmitted one picture of the surface.
- Pioneer Venus 1 (USA) launched 20 May 1978 arrived 4 December 1978. Successful orbiter.
- Pioneer Venus 2 (USA) launched 8 August 1978 arrived 4 December 1978. 1 of 4 probes landed safely. Other three failed upon landing.
- Venera 11 (USSR) launched 9 September 1978 arrived 25 December 1978. Survived for 95 minutes after landing.
- Venera 12 (USSR) launched 14 September 1978 arrived 22 December 1978. Survived for 60 minutes after landing.
- Venera 13 (USSR) launched 30 October 1981 arrived 1 March 1982. Analysed Venusian soil.
- Venera 14 (USSR) launched 4 November 1981 arrived 5 March 1982. Analysed Venusian soil.
- Venera 15 (USSR) launched 2 June 1983 arrived 10 October 1983. Radar mapper.
- Vega 1 (USSR) launched 15 December 1984 arrived 11 June 1985. Dropped lander and balloon en route to Halley's Comet.
- Vega 2 (USSR) launched 20 December 1984 arrived 15 June 1985. Dropped lander and balloon en route to Halley's Comet.
- Magellan (USA) launched 5 May 1989 arrived 10 August 1990. Radar mapper.
- Galileo (USA) launched 18 October 1989 arrived 10 February 1990. Passed Venus en route to Jupiter.

MISSIONS TO MARS

- Mars 1 (USSR) launched 1 November 1962. Failed 106,000 km (65,868 miles) from Earth.
- Mariner 3 (USA) launched 5 November 1964. Contact lost after launch.
- Mariner 4 (USA) launched 28 November 1964 arrived 14 July 1965. Took 21 pictures during flyby.
- Zond 2 (USSR) launched 30 November 1964. Contact lost five months after launch.
- Mariner 6 (USA) launched 24 February 1969 arrived 31 July 1969. Took 76 pictures during flyby.
- Mariner 7 (USA) launched 27 March 1969 arrived 4 August 1969. Took 126 pictures during flyby..
- Mars 2 (USSR) launched 19 May 1971 arrived 27 November 1971. Landed but returned no images.
- Mars 3 (USSR) launched 28 May 1971 arrived 2 December 1971. Partially successful orbiter, lander failed.
- Mariner 9 (USA) launched 30 May 1971 arrived 13 November 1971. Successful orbiter – took 7,329 images.
- Mars 4 (USSR) launched 21 Jul 1973 arrived 10 February 1974. Failed to orbit – returned some data during flyby.
- Mars 5 (USSR) launched 25 July 1973 arrived 12 February 1974. After returning 60 images of Mars, transmitter failed.
- Mars 6 (USSR) launched 5 August 1973 arrived 12 March 1974. Contact lost during landing.
- Viking 1 (USA) launched 20 August 1975 arrived 19 June 1976. Successful orbiter and lander.
- Viking 2 (USA) launched 9 September 1975 arrived 7 August 1976. Successful orbiter and lander.
- Phobos 1 (USSR) launched 7 July 1988. Contact lost en route to Mars.
- Phobos 2 (USSR) launched 12 July 1988. Contact lost while approaching Mars.
- Mars Observer (USA) launched 25 September 1992. Contact lost while approaching Mars.
- Mars Global Surveyor (USA) launched 7 November 1996 arrived 11 September 1997. Successful orbiter.
- Mars 96 (former USSR) launched 16 November 1996. Failed during launch.
- Mars Pathfinder (USA) launched 4 December 1996 arrived 4 July 1997.

Successful lander and rover.
- Nozomi (Japan) launched 3 July 1998. Failed upon reaching Mars.
- Mars Climate Orbiter (USA) launched 11 December 1998. Failed upon reaching Mars.
- Mars Polar Lander (USA) launched 3 Jan 1999. Failed upon reaching Mars.
- Mars Odyssey (USA) launched 7 April 2001 arrived 24 October. Successful orbiter.
- Mars Express (Europe) launched 2 June 2003 arrived 25 December. Successful orbiter; lander (Beagle 2) failed upon reaching Mars.
- Spirit launched 10 June 2003 and Opportunity (both USA) launched 7 July 2003. Arrived mid-January 2004; successful landers and rovers.

MISSIONS TO THE OUTER SOLAR SYSTEM

- Pioneer 10 (USA) launched 2 March 1972. First flyby of Jupiter, 3 Dec 1973
- Pioneer 11 (USA) launched 5 April 1973. Jupiter flyby, 2 December 1974. Went on to fly by Saturn, 1 September 1979.
- Voyager 2 (USA) launched 20 August 1977. Visited Jupiter (1979), Saturn (1981), Uranus (1986) and Neptune (1989).
- Voyager 1 (USA) launched 5 September 1977. Visited Jupiter (1979) and Saturn (1980).
- Galileo (USA) launched 18 October 1989. Passed by Venus and two asteroids en route to Jupiter. Began orbiting Jupiter 7 September 1995 to begin extensive survey. Still in operation.
- Ulysses (USA) launched 6 October 1990. Jupiter flyby, 8 February 1992, to establish a polar orbit around the Sun.
- Cassini-Huygens (USA/Europe) launched 15 October 1997. Began extensive survey of Saturn and its moons in 2004.

MISSIONS TO THE NEAR SOLAR SYSTEM

- Giotto (Europe) launched 2 July 1985. Passed within 596 km (370 miles) of the nucleus of Halley's Comet on 13 March 1986.
- Stardust (USA) launched 7 February 1999. Successfully encountered comet Wild 2 in January 2004. Samples of interplanetary dust will be returned to Earth in 2006.
- Rosetta (Europe) launched 2 March 2004. Due to rendezvous with comet Wirtanen in 2011.

Glossary

Apollo: Ambitious and successful US series of manned lunar missions. Eleven manned missions in all, six of which landed on the Moon.

Asteroid: One of countless rocky bodies or 'minor planets' in orbit around the Sun.

Astrology: The forerunner of astronomy in which astrologers attempted to use the motions of the planets to predict the future.

Astronaut: 'Star voyager' – highly trained person who travels to space. Traditionally used to describe US personnel. Cosmonaut is Soviet/ Russian equivalent.

Astronomy: Scientific discipline dealing with everything beyond the Earth's atmosphere.

Astrophotography: Specialist photography focussing on astronomical objects and celestial bodies.

Atmosphere: Shell of gases surrounding a planetary body.

Atom: The smallest component of an element which exhibits the properties of that element. Consists of protons, neutrons and electrons.

Axis: Imaginary line drawn through the middle of a rotating body, such as the Earth, around which it rotates.

Background Radiation: Extremely faint energy coming from every direction in the Universe. Believed to be a ghostly echo of the Big Bang.

Big Bang: Accepted by most astronomers to have been the origin of the Universe around a 12 billion years ago.

Binary Star: Star system in which two stars orbit each other, in the same way that the Moon orbits the Earth.

Black Hole: Region surrounding a tiny, super-dense remnant of a collapsed, dead star, with a gravitational field so high that not even light can escape or reflect from it.

Comet: Small icy body, usually a few kilometres across, originating in the outer Solar System. If they approach the Sun, the ice begins to evaporate and streams away in a tail, millions of kilometres long.

Constellation: Apparent visible grouping together of stars in the sky, usually named after people, animals, gods or objects.

Cosmonaut:. Term used to describe highly trained Soviet/Russian person who travels into space. Same meaning as 'astronaut'.

Crater: Bowl-shaped depression on a planetary body. Caused either by violent impacts or volcanic explosions.

Eclipse, Lunar: When the Moon passes into the shadow of the Earth and light from the Sun is cut off.

Eclipse, Solar: When the Moon passes directly between the Earth and the Sun and projects its shadow on to the Earth.

Electromagnetic Spectrum: The full range of radiation in the Universe, from long-wavelength radio waves, through visible light and on to gamma rays, the most energetic.

Elliptical orbit: All bodies orbit in ellipses, which are oval-shaped. A circle is a perfect ellipse.

Epicycle: A moving circle about which a planet orbits, proposed by Ptolemy, to explain the motion of the planets in the sky.

ESA: The European Space Agency – a co-operative organisation consisting of 14 European nations.

Galaxy: A vast system of stars, gas and dust. Many are spiral shaped, like our own galaxy, the Milky Way.

Gas Giant: A type of planet, with no solid surface, much larger than the rocky planets, such as Earth. Jupiter, Saturn, Uranus and Neptune are all gas giants.

Gemini: Successful series of US manned space missions from 1964 to 1966 – the forerunner to the Apollo programme.

GPS: Global Positioning System – a series of Earth-orbiting satellites used for navigation.

Gravity: The physical force which causes matter to attract other matter.

Infrared: A region of the electromagnetic spectrum with longer wavelengths than visible light. We feel infrared radiation on Earth as heat.

Interferometry: An astronomical technique which combines the light/radiation collected by more than one telescope to make images better than

one telescope could produce by itself.

Ion Drive: Experimental type of spacecraft propulsion much more efficient than conventional rocket propulsion. Currently being evaluated by NASA's Deep Space 1 probe.

ISS: International Space Station, due for completion in the next few years. Designed to be an Earth-orbiting manned facility in which to perform scientific experiments of numerous types and also promote and further international scientific co-operation.

Latitude: Imaginary lines crossing the Earth, parallel to the equator. Used for navigation.

Light Year: The distance light travels in one year, in a vacuum. Equivalent to almost 9.5 million million kilometres (5.8 million million miles).

Longitude: Imaginary lines crossing the Earth and used for navigation. They form a grid system with the lines of latitude.

Lunar Orbit: Used to describe the elliptical path a spacecraft follows when circling the Moon.

Manned Mission: A mission into space carried

out by astronauts/ cosmonauts in a spacecraft.

Mercury Programme: The first series of US manned missions from 1961 to 1963. There were three unmanned tests and six piloted missions – the forerunner to the Gemini programme.

Meteors: Small natural particles in space, usually sand grain-sized. Visible when they enter the Earth's atmosphere and burn up due to friction. Most originate in the tails of comets.

MGS: Mars Global Surveyor. US unmanned mission to Mars, launched in 1996 and still operational. Mission is detailed mapping of the surface from orbit.

Microwave: Type of radiation in the electromagnetic spectrum. Longer wavelength than infrared and shorter than radio waves.

Milky Way: Faint band of light crossing the sky, a view of our spiral galaxy edge-on and from within. The light represents the combined light from millions of distant stars. Also used as the name for our Galaxy.

Module: Section of spacecraft/space station, usually with its own power supply and life-support systems.

NASA: National Aeronautics and Space Administration. American space agency established in 1958 by President Eisenhower with the goal of catching up with Soviet space achievements.

Nebula: Diffuse cloud of gas and/or dust in space. Can be formed by dying stars or can be raw material from which stars and planets condense and form.

Neutron Star: Remnant left behind when a massive star explodes in a supernova. Contains the mass of a star compressed into a sphere the size of a city.

Nucleus: Used to describe the centre of atoms, galaxies, comets and other physical phenomena.

Orbit: The path followed by one celestial body around another, prevented from flying off into space by gravity.

Orrery: Often beautiful clockwork model of the Solar System, demonstrating the orbits of the planets around the Sun at the correct relative speeds.

Parallax: The apparent movement of an object against the background, when seen from different directions.

Phase: The changing appearance of the Moon over the course of its orbit around the Earth, due to how much of its surface is illuminated by the Sun. Ranges from 'full Moon' when fully lit, to 'new Moon' when just the slimmest crescent is visible.

Photosphere: Essentially the surface of the gaseous Sun, from which most of the light escapes into outer space.

Planet: 'Wandering star', the major natural satellites of the Sun, of which there are nine, including the Earth. Recent advances in astronomy have led to the discovery of planets orbiting other stars.

Probe: Unmanned spacecraft used for exploration and armed with a variety of scientific detectors, usually including cameras.

Protostar: An early stage in the formation of a star.

Pulsar: A type of neutron star, which rotates extremely quickly, detectable by radio astronomy as beams of radiation from the pulsar

sweep over the Earth.

Radiation: Energy which permeates the Universe at the speed of light. The range of types and strengths of radiation is called the electromagnetic spectrum.

Radio Waves: Low energy, long wavelength region of the electromagnetic spectrum. Unlike visible light, radio waves can easily penetrate dust and gas clouds in the Universe, making them a valuable tool for astronomers.

Red Dwarf: Low mass, cool, small star with surface temperature around half of that of the Sun.

Retrograde Orbit: Path followed around a celestial object in the opposite direction to which the object rotates. For example, some of the irregular, small, outer moons of Jupiter, and Triton (moon of Neptune).

Rocket: A vehicle powered by explosive propellants, which create lift and allow the rocket to travel above the Earth's atmosphere.

Satellite: Relatively small planetary body, such as the moons of the planets, which orbit larger celestial objects, or manmade objects, like weather satellites, that orbit the Earth.

Salyut: Series of seven Soviet space stations beginning in 1971.

Shooting Star: Popular name for a meteor, as it burns up in the Earth's atmosphere.

SOHO: European Space Agency's Solar and Heliospheric Observatory, launched in 1995.

Solar System: Everything under the direct influence of the Sun. Used to describe the Sun and planets, as well as their moons and the asteroids and comets.

Soyuz: Series of Soviet/ Russian manned space missions. Over 80 of these spacecraft have been launched since 1967.

Space Race: The competition between America and the Soviet Union in the 1960s to put a man on the Moon.

Space Shuttle: The latest American manned spacecraft. NASA has four delta-winged, re-usable shuttles, the first of which, Columbia, was launched in 1981 and is still in service.

Spacewalk: Also known as EVA 'Extra Vehicular Activity', when an astronaut/cosmonaut leaves the protection of his spacecraft in a spacesuit in order to work outside in the vacuum of space.

Space Debris: Manmade junk orbiting the Earth and often presenting a hazard to space missions.

Speed of Light: The velocity travelled by radiation, equal to around 300,000 kps (186,420 mps). According to current theory, nothing in the Universe can exceed this speed.

Star: A sphere of gas in space, like the Sun, which shines due to nuclear reactions in its core.

Star Cluster: A group of stars close together in space, which formed together from the same nebula.

Sunspot: A dark blemish on the surface of the Sun caused by magnetic fields. They appear dark because they are cooler than the rest of the Sun's surface.

Supergiant: An extraordinarily large, luminous star.

Supernova: A titanic explosion which occurs when a supergiant star collapses then explodes at the end of its life.

Telescope: Instrument for magnifying portions of the sky and viewing remote objects.

Totality: A period during either a lunar or solar eclipse when the Moon or the Sun is completely obscured.

Ultraviolet radiation: A region of the electromagnetic spectrum with shorter wavelengths than visible light. Most ultraviolet radiation from the Sun is blocked by Earth's ozone layer in the upper atmosphere, with the remainder reaching the ground and causing sunburn in humans.

Universe: Everything that was created in the Big Bang, including galaxies, stars and planets.

Vacuum: The complete lack of matter in a given volume.

Voyager: Two space probes which, between them, explored Jupiter, Saturn, Uranus and Neptune between 1979 and 1989.

Weightlessness: The lack or apparent lack of gravity.

White Dwarf: A remnant of a dead Sun-like star after the star has blown away its atmosphere.

Wormhole: A theoretical conduit that can link one region of the Universe with another.

Useful Information

FURTHER READING

Beatty, J. Kelly, Collins Petersen, Carolyn and Chaikin, Andrew, *The New Solar System* (Cambridge University Press, 1990)

Burnham, Robert et al, *Spacewatching* (HarperCollins, 1998)

Hawking, Stephen, *A Brief History of Time* (Bantam, 1995)

Moore, Patrick, *Atlas of the Universe* (Cambridge University Press, 1998)

Moore, Patrick, *Patrick Moore On Mars* (Cassell, 1998)

Moore, Patrick, *The 2001 Yearbook of Astronomy* (Pan, 2000)

Nicolson, Iain, *Unfolding our Universe* (Cambridge University Press, 1999)

Ridpath, Ian, *Night Sky* (HarperCollins, 1999)

Sagan, Carl, *Cosmos* (Abacus, 1995)

WEBSITES

http://www.nasa.gov The NASA homepage
http://www.iki.rssi.ru The Russian Space Agency homepage
http://www.sci.esa.int The ESA science homepage
http://www.planetary.org The Planetary Society
http://www.seds.lpl.arizona.edu Students for the exploration and development of space
http://www.stsci.edu/ Hubble Space Telescope information
http://www.astwrp.gsfc.nasa.gov/apod/astropix.html Astronomy picture of the day
http://www.hq.nasa.gov/office/pao/NewsRoom/History/Whatsnew.html What's new in NASA history and online information
http://www.obspm.fr/departement/darc/planets/encycl.html Extra-solar planets
http://sohowww.nascom.nasa.gov/ SOHO mission
http://www.astro.ucla.edu/~obs/intro.html Pictures of the Sun
http://nssdc.gsfc.nasa.gov//planetary/lunar/ Lunar exploration
http://www.jpl.nasa.gov/galileo Galileo mission information
http://www.jpl.nasa.gov/cassini Information on the Cassini mission to Saturn
http://www.hq.nasa.gov/osf/heds/ Information on human spaceflight

ACKNOWLEDGEMENTS

The author would like to thank Polly Willis and Sonya Newland at The Foundry for their patience and encouragement. Special thanks to Lara Speicher for her unending help and support. This book is dedicated to Sam and Glad Hawksett, and to Shandy.

Grateful thanks also to Carole Stott.

PICTURE CREDITS

Index